Cambridge
checkp•int

SECOND EDITION

Lower Secondary
Science

REVISION GUIDE

FOR THE SECONDARY 1 TEST

**Rosemary Feasey, David Bailey and
Andrea Mapplebeck**

HODDER
EDUCATION
AN HACHETTE UK COMPANY

Photo credits
p.9 © Sara Sadler/Alamy Stock Photo; **p.21** © Xinhua/Shutterstock; **p.23** © Steve McHale/stock.adobe.com, *t & br*; **p.23** © Galyna/stock.adobe.com, *bl*; **p.27** © Eric Isselée/stock.adobe.com; **p.31** © James King-Holmes/ Science Photo Library, *t*; **p.31** © BVpix/stock.adobe.com, *b*; **p.32** © Sheila Terry/Science Photo Library; **p.33** © Classic Image/Alamy Stock Photo, *tl & tr*; **p.33** © Ronstik/stock.adobe.com, *bl*; **p.33** © Maridav/stock.adobe. com, *br*; **p.43** © Nenov Brothers/stock.adobe.com; **p.47** © Kay & Karl Ammann/Avalon/Bruce Coleman Inc/ Alamy Stock Photo; **p.54** © Prochym/stock.adobe.com, *tl*; **p.54** © Ch.krueger/stock.adobe.com, *tr*; **p.54** © Warren/stock.adobe.com, *bl*; **p.54** © Vkilikov/stock.adobe.com, *br*; **p.60** © INTERFOTO/Personalities/Alamy Stock Photo; **p.68** © David Bailey (author); **p.86** © Andrew Lambert Photography/Science Photo Library; **p.92** © Andrew Lambert Photography/Science Photo Library; **p.95** © Martyn F. Chillmaid; **p.110** © VectorMine/stock. adobe.com; **p.118** © Freedom Life/stock.adobe.com; **p.119** © Anitalvdb/stock.adobe.com.

Orders: please contact Hachette UK Distribution, Hely Hutchinson Centre, Milton Road, Didcot, Oxfordshire, OX11 7HH. Telephone: +44 (0)1235 827827. Email education@hachette.co.uk Lines are open from 9 a.m. to 5 p.m., Monday to Friday. You can also order through our website: www.hoddereducation.com

ISBN: 9781398364219

First published in 2013
This edition published in 2023 by
Hodder Education,
An Hachette UK Company
Carmelite House
50 Victoria Embankment
London EC4Y 0DZ

www.hoddereducation.com

Impression number   10 9 8 7 6 5 4 3

Year   2026 2025 2024

Cover photo © NicoElNino - stock.adobe.com

Illustrations by Integra Software Services Pvt. Ltd., Pondicherry, India

Typeset by Integra Software Services Pvt. Ltd., Pondicherry, India

Printed by Ashford Colour Press Ltd

A catalogue record for this title is available from the British Library.

Contents

Biology

Chemistry

Physics

Earth and space

Introduction

What is this book about?

This Revision Guide aims to help you to recall and remember key information and ideas and build your understanding about the science topics that you have been learning during Stages 7–9. Revision helps you make connections between these ideas so the knowledge and understanding you develop stay in your memory for longer.

Where can you use this book?

You can work through this book on your own or with someone else at school or at home, or both. You can write in the book but there will be times when you need to use a separate piece of paper to work.

How can you help yourself when using this book?

Be honest with yourself; if you do not know something or are unsure, it is fine to admit it. If you still do not understand an idea or a word when revising, ask a friend or an adult to help you understand. Positive learners know that asking for help is a good thing to do.

If you are completing revision at home, try to find a quiet area, and make sure the space you are working in is tidy before you begin. A tidy desk helps to create a calm area with plenty of room to work and no distractions.

When you have finished a revision section, reward yourself: enjoy your favourite treat!

Features in this book

Throughout the book, certain features are used to guide you in using the revision material.

CHAPTER INFORMATION
These are the key ideas that you will be learning in each chapter.

REMEMBER
These sections provide information to remind you of what you have learned.

Activity

These are activities to carry out, to help you put into practice what you have learned. You can either put your answers in this book, or in a notebook.

TAKE A BREAK

This indicates a point where you should take time out from working on your revision activities.

REVISION APPROACHES

This shows where specific approaches to revision are used and explains what they are. These revision approaches will help you to recall ideas and information and deepen your learning.

Revision test

There is a revision test at the end of each chapter, covering learning from that chapter, so that you can practise for tests and exams.

TIPS FOR SUCCESS

These suggest things to think about and do before your revision test.

MODEL ANSWER

A model answer is an 'ideal' response to a question. Looking at different answers to questions and thinking about how they can be improved will help you to revise and remember your science.

Chapter 1 Water and life

CHAPTER INFORMATION

This chapter will help you to revise your learning about how water is important to plants, how water is transported in plants, and about the human renal system. By the end of this chapter you should be able to:

- describe the pathway of water and mineral salts from the roots to the leaves in flowering plants
- explain how water is absorbed in root hair cells
- describe how water is transported through plants in xylem vessels
- describe how water that evaporates from the leaves is replaced by water from the xylem vessels in a process called transpiration
- explain how water absorbed from the soil through osmosis contains mineral salts that are used by plants
- describe the structure of the human excretory (renal) system and how the kidneys remove urea from the blood
- describe how plants use magnesium and nitrates.

REVISION APPROACH: PLANTS KEY WORD CONCEPT MAP

You are going to begin this chapter by using a revision strategy called a concept map. Concept maps are useful for helping you to remember ideas and make connections between those ideas. They enable you to see what you know and make links across a topic.

By creating a concept map, you will be able to see the bigger picture of how to link ideas about minerals and how plants take in water. Concept maps also help you to 'dig deeper' into your memory, because putting one thing on the concept map can lead to your memory linking to another idea.

You are going to begin by making key word cards to use when you create your concept map.

Activity 1: Key word cards

a) Look at the list of words below. Which words do you already know? Highlight these words.

chlorophyll	nitrate	protein	vacuole
leaf	osmosis	root	vascular bundle
magnesium	photosynthesis	stem	xylem
mineral salt	plant	transpiration	xylem vessel

b) Make key word cards for all the words you highlighted. Write each word on the front of a piece of card and add any images that help you remember what the word means.

c) On the back of each card, write as much information as you can linked to the word. This could include a definition, facts or examples. Do not worry if you do not know all the words yet; you can make cards later for the new words you learn. You will repeat this activity at the end of this chapter when you know more words.

Front of card	Back of card
Stem	Stem
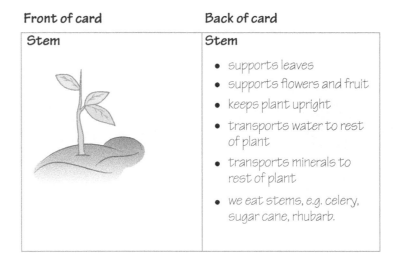	• supports leaves • supports flowers and fruit • keeps plant upright • transports water to rest of plant • transports minerals to rest of plant • we eat stems, e.g. celery, sugar cane, rhubarb.

Activity 2: Plants concept map

a) Using the key word cards you made in Activity 1, sort the cards you know into groups; for example, everything to do with leaves.

b) Lay out the key word cards you have made on a large sheet of plain paper.

c) Now draw lines to join as many of the cards as you can on the sheet, and on the connecting lines write the reason why you have linked the cards. The more lines you draw, the more links there are that show your knowledge and understanding. The picture below shows a concept map that has just been started.

d) When you have finished, take a photograph of your concept map. Then collect up your word cards and keep them safe. You will need them again at the end of this chapter.

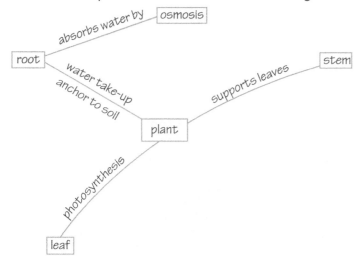

REMEMBER: MOVEMENT OF WATER THROUGH A FLOWERING PLANT

In flowering plants, water and dissolved minerals from the soil enter the plant through the roots. Roots have root hairs; these give the root system a greater surface area, allowing the plant to take up more water. The root hairs are numerous, long and thin, which allows them to get between soil particles to reach water.

Water enters the roots by a process called osmosis. Osmosis is the movement of water molecules across a partially permeable membrane, from an area of higher water concentration to an area of lower water concentration. In plants, the cell membrane allows water and minerals to pass through and so allows water to be transported around the plant.

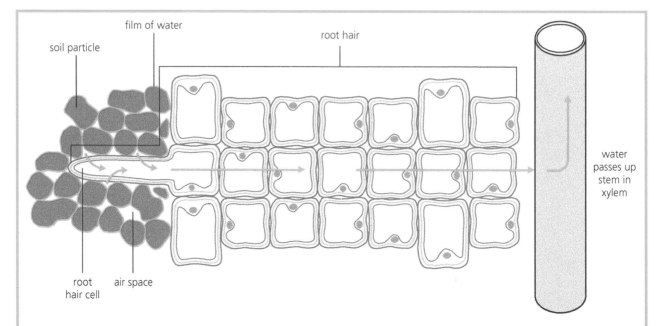

Plants need minerals for healthy growth. Minerals dissolved in the water are taken up by the roots and absorbed by the plant. Two important minerals that plants need are nitrogen and magnesium. These help to make chlorophyll. Without nitrogen, plants cannot make chlorophyll and the leaves turn yellow. The plant then shows poor growth.

The movement of water through a flowering plant is controlled by the process of transpiration. Transpiration is the loss of water through evaporation, mostly from the lower side of the leaves. The movement of water from the roots, through the stem to the leaves is called the transpiration stream. Follow this movement of water in the diagram below.

As a plant transpires and water evaporates from the lower side of the leaves, the cells in the lower layer of the leaf become short of water. These cells then take in water from the xylem to replace the water that has been lost by evaporation. As the xylem loses water to the leaves, water is then drawn in by the plant using the roots.

Water and minerals therefore pass from the root hairs through the xylem and into the leaves and all other parts of the plant. In flowering plants, most of the cells that make up the xylem are called vessels. These vessels do not have ends to their walls, so the xylem forms a continuous, hollow tube that water and minerals can flow through.

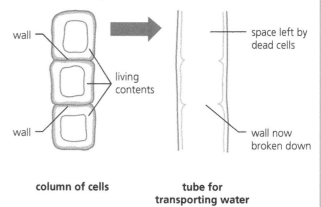

Activity 3: Movement of water through celery

Look at the photograph below.

a) Describe what the leaves show about the movement of water through the celery.

....

....

....

b) Describe what the cross-section through the celery stem shows.

....

....

c) Explain how the cross-section through the celery stem helps to show the movement of water through a flowering plant.

....

....

Activity 4: Removing leaves

Learners placed two celery plants with a similar number of leaves in one container of blue water and one container of red water, as shown in the picture in Activity 3. The volume of the blue water and red water was the same. They then took all the leaves off the celery in the red water, and left all the leaves on the celery in the blue water.

Predict how taking the leaves off the celery in the red water would affect the rate at which water was transported in that celery stem.

....

....

....

Activity 5: Annotated diagram

Draw and label a diagram to show how water is transported in a plant.

TAKE A BREAK

Now take a short break from revising. Go and drink one or two glasses of water before you begin the next section.

Scientists have researched the effect of learners drinking water when they take exams. Learners who drank water regularly did about 5 per cent better than those who did not. Staying hydrated can help your memory and concentration.

Activity 6: Revisit your concept map

a) Use the concept map word cards that you made at the start of this chapter in Activity 1.

b) Highlight any words from the list in Activity 1 that you now know but did not at the beginning, and make new cards for those words.

c) Group the cards on the plain paper to create a new concept map using all the cards.

d) Draw lines to join as many of the cards as you can, making sure that you write what connects the cards on each line.

e) Compare this concept map to the one you made at the start of the unit. How many new connections can you now make?

f) Give your concept map to a partner and ask them to pick two words that you have connected. You must then explain to them how you have connected these words.

g) Take a photo of your concept map. Stick the corner of each card down so that you can still read the information on the back. Use your map for revision to help you remember the connections you have made.

REMEMBER: THE HUMAN RENAL SYSTEM

Taking a break to drink water is a good place to start this section of revision, because it is about the human renal system. Anything using the word 'renal' has something to do with the kidneys. The kidneys filter your blood, removing waste and excess water.

It is important that the body keeps its water levels steady. We take in water through food and drink, and lose water through the skin by sweating and when we breathe out (exhale). We also lose water when we urinate (go to the toilet). In this section, we will focus on the role of the kidneys in keeping the level of water in our body constant.

Humans have two kidneys, one on either side of the back above the waistline. The kidneys are about the size of an adult fist.

Activity 7: Five key facts

Choose five facts from the Remember section on the human renal system that you think a learner should know.

1 ....

2 ....

3 ....

4 ....

5 ....

REMEMBER: HOW THE KIDNEYS WORK

The kidneys have two important functions:

- to control the amount of water in the body
- to remove waste (urea) from the body.

The body breaks down food for energy and to repair itself. Any waste from this process enters the blood. The main function of the kidneys is to remove this waste from the blood and return the cleaned blood back to the body. The blood enters each kidney through the renal arteries, which are two large blood vessels. Each renal artery then branches into smaller and smaller blood vessels and finally into tubes called nephrons, which act as tiny filters. Each kidney contains about 1 million of

these microscopic nephrons. Blood passes through the nephrons and is filtered to remove a substance called urea, which is a toxin that the body needs to get rid of. After the blood is cleaned, it flows out of the kidneys through the renal veins.

The waste from the blood, together with any excess water, is known as urine. The urine flows from the kidneys to the bladder through the ureters. The ureter is a tube from each kidney to the bladder, so there is a ureter on each side of the bladder. The urine is stored in the bladder and then passes from the body through a tube called the urethra when you go to the toilet.

Activity 8: Label the diagram

Use the information from the Remember section to label the diagram below.

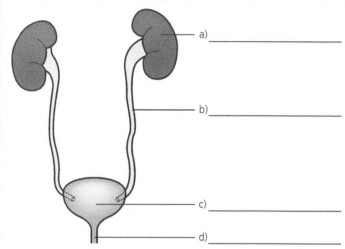

a) _____

b) _____

c) _____

d) _____

Activity 9: Parts of the kidney

Read the Remember section on how the kidneys work to help you complete this activity. Draw lines to match each scientific term related to the renal system to the correct definition.

nephron	Carries blood from your heart to your kidneys
ureter	A filter that removes urine
renal vein	Carries cleaned blood back to the heart
renal artery	A tube from each kidney that carries urine to the bladder

Learners were challenged to make a working model of the renal system, which would show how the system filters blood to remove urea and create urine. Below are three pictures of different models made by learners. Look carefully at each model and reflect on how the learners have represented the different parts of the system. Then answer the questions below.

Remember that a model represents something; it is not the real thing, so it may have limitations and not work exactly as a real renal system would.

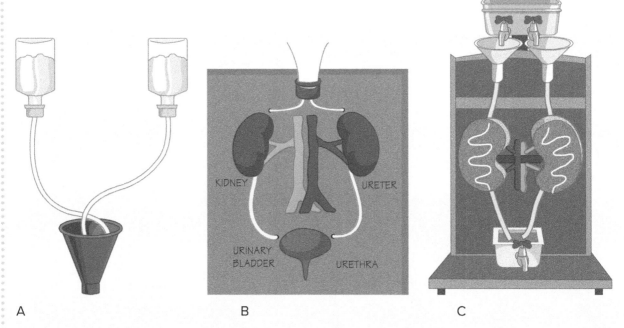

A   B   C

List the different parts of the renal system shown in each model in the following table, and say how each part is represented in the model. One has already been started for you.

Model	Part of the renal system	How it is represented in the model
A	Kidneys Bladder	Plastic bottles with filter paper Filter funnel
B		
C		

Activity 11: PMIs

What are the PMIs (**P**ositives, **M**inuses and **I**nteresting features) for each model shown in Activity 10? Complete the table to show your ideas. The first one has been done for you.

Model	Positives	Minuses	Interesting features
A	Shows the main parts of the system (kidneys, ureters and bladder)	Less detail than the other models	The way filter paper has been used to show a filtering system
B			
C			

Activity 12: Best model

Which model do you think will best help learners to understand the renal system? Explain why, using your knowledge of the renal system.

....

....

....

TIPS FOR SUCCESS

Go back over the work you have done in this chapter to remind yourself of all the information you have covered. When you are ready, complete this short test.

As you work through it, you can help yourself by:

- reading each question carefully – check you understand the question
- looking for key words to use in your answer
- answering the question in your mind first, before you write it down
- making sure you use correct scientific vocabulary in your answers
- using a piece of spare paper to draft any extended answers first, then when you are happy with it you can write your answer in this book
- checking your answers to make sure that you do not want to make any changes.

Revision test

1 Which of the following words is part of the renal system? Circle the correct answer.　[1 mark]

　nephron　root hair　transpiration　vascular bundle　xylem

2 Which of the following could be shown by monitoring the colour of urine? Circle the correct answer.　[1 mark]

　A　How much oxygen is in the blood

　B　Whether someone is hydrated

　C　Whether someone has a healthy diet

3 Complete the following table by ticking the correct column to show whether each statement is true or false.　[5 marks]

Statement	True	False
Roots absorb water and dissolved minerals.		
Transpiration is when a plant takes in water from the air.		
Root hairs have a large surface area for absorption of water.		
Most of the water lost by plants evaporates from the stem.		
Xylem transports water and dissolved minerals from the roots to the rest of the plant.		

4 Write definitions for each of the following:　[3 marks]

　a) Xylem ....

　...

　b) Nephron ....

　...

　c) Urea..

　...

5 Describe the difference in the functions of the renal arteries and the renal veins.　[2 marks]

　...

　...

　...

6 The title of this chapter is 'Water and life'. Suggest why both flowering plants and human kidneys are included together in this chapter.　[3 marks]

　...

　...

　...

　...

　...

7 a) Draw arrows on the diagram to show the transpiration stream in the plant.   [1 mark]

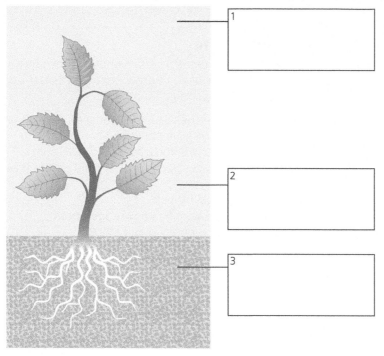

1 [   ]

2 [   ]

3 [   ]

b) Add labels to the diagram to describe the processes in the transpiration stream shown at points 1, 2 and 3.   [3 marks]

8 Label the parts of the renal system represented in this model.   [4 marks]

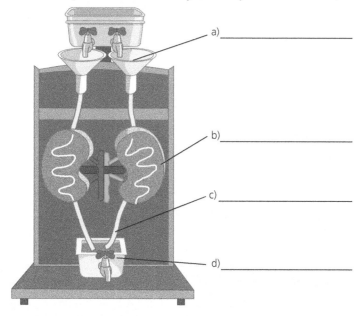

a)_____

b)_____

c)_____

d)_____

Chapter 2 Photosynthesis

REMEMBER: SCIENTISTS AND OUR UNDERSTANDING OF PLANTS

Over the last 450 years, scientists have gradually built upon each other's work in order to understand how plants grow. Joannes Baptista van Helmont (1580–1644) carried out an experiment where he watered a willow tree and measured its mass, which led him to believe that plants need only water to grow. We now know that this was incorrect, but it led to further work by other scientists who wanted to check his idea. Stephen Hales (1677–1761) thought that 'a portion of air' helped a plant to survive. Jan Ingenhousz (1730–1799) built on the knowledge of van Helmont and Hales to show that the 'portion of air' that green plants take up is carbon dioxide when they are in the sunlight. Later, Joseph Priestly (1733–1804) placed a mint plant in a jar of this gas and let sunlight shine on it. He found that the gas appeared to change into another one that allowed things to burn in it. Later, Priestley met the French chemist Antoine Lavoisier (1743–1794) and told him about his discovery. Lavoisier named the gas oxygen.

Activity 1: Timeline

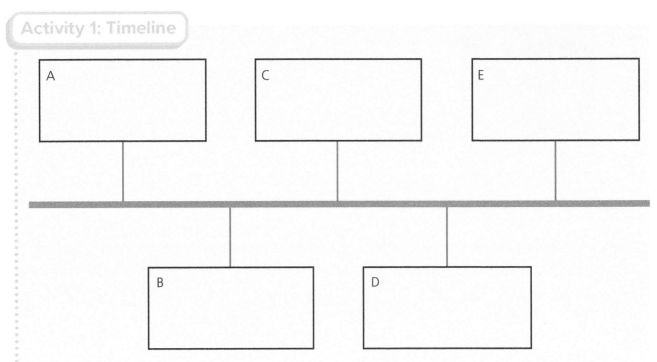

Above is a timeline with five empty boxes. Use the Remember section to help you put each scientist in the correct place on the timeline, to show when they made their contributions.

Activity 2: Matching scientists to ideas

Draw a line to match each scientist with the scientific idea they developed.

Scientists	Ideas
Joseph Priestley	Water is a basic requirement for life.
Antoine Lavoisier	The gas produced by plants is named oxygen.
Jan Ingenhousz	A gas produced by plants allows things to burn in it.
Stephen Hales	Air helps plants to survive.
Joannes Baptista van Helmont	Plants take up carbon dioxide.

Activity 3: Using information

a) What did Joannes Baptista van Helmont investigate? How is his contribution to science important to our knowledge of plants today?

....

....

....

b) How did Stephen Hales' work contribute to scientists' understanding of plant growth?

....

....

....

c) How did Jan Ingenhousz's work add to knowledge about plant growth?

....

....

....

d) Joseph Priestley and Antoine Lavoisier met and shared their scientific findings. What did they discuss and what gas did Lavoisier name?

....

....

....

e) Why are plants and photosynthesis crucial to life on Earth? Explain your answer.

....

....

REMEMBER: TESTING FOR STARCH

Scientists use a range of tests when working with plants. One is called the starch test, which is used to look for the presence of starch to find out if photosynthesis has occurred.

Plants use energy from the Sun to change water and carbon dioxide into a sugar called glucose. This can be expressed in the word equation:

carbon dioxide + water → glucose + oxygen

Glucose is a simple sugar that is used by plants for energy and to make other substances, like cellulose and starch, which are used to form the plant's structure. Starch is made up of glucose units linked together; plants make and store starch and then break it down into glucose when they need energy.

So, photosynthesis needs sunlight, carbon dioxide and water to produce glucose.

Activity 4: Starch test

a) Stage 9 learners were planning to test a leaf for starch. In this diagram, they are preparing the leaf for the starch test. Complete the diagram by adding the labels.

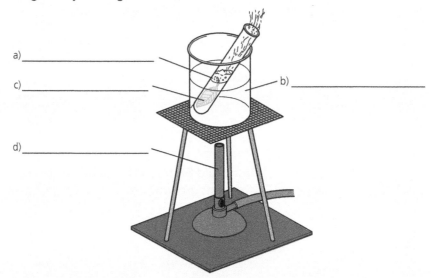

a)_____

c)_____

b) _____

d)_____

b) What effect does ethanol have on the leaf?

....

....

c) Which substance in a leaf captures light for energy? ....

d) What colour will the leaf turn if it contains starch? ....

e) Which sugar do plants make when they photosynthesise?..........................

f) What happens in the chloroplasts in a leaf?

....

REVISION APPROACH: MODEL ANSWER

A model answer is an 'ideal' response to a question. Looking at different answers to questions and thinking about how they can be improved is one way of helping you to revise and remember your science. Thinking about the strengths and weaknesses of an answer helps you assess how good the answer is, and rewriting it to improve the answer can help you later when you have to write an answer yourself.

Activity 5: Criteria for model answer

Write a set of four criteria that a teacher could use to mark an answer to this question:

Describe how to carry out a starch test.

1 ....

2 ....

3 ....

4 ....

Activity 6: The role of carbon dioxide in photosynthesis

Learners were testing Jan Ingenhousz's hypothesis that green plants take up carbon dioxide from the air when they are put in sunlight. They investigated the effect of carbon dioxide on starch production in leaves.

Two de-starched plants were placed in individual transparent plastic bags. One was enclosed with a dish of soda lime, and the other with a dish of sodium hydrogencarbonate solution. The two plastic bags were sealed and both plants were left in a sunny place for a few hours.

plastic bag

soda lime absorbs carbon dioxide from the air

sodium hydrogencarbonate solution releases carbon dioxide into the air

a) Why did the learners use de-starched plants?

....

b) List three things that the learners should control to make sure their test is fair.

1 ....

2 ....

3 ....

c) Why did the learners place both plants in a sunny place instead of a dark cupboard?

....

d) What will the learners need to do to find out the effect of carbon dioxide on each plant?

....

e) Predict the outcome of this investigation and explain the reasons for your prediction.

....

....

....

Activity 7: Dual coding

Dual coding uses images alongside words to help you remember complex ideas; for example, remembering equations such as the one for photosynthesis.

Look at this part of the photosynthesis equation. Underneath the words, add some drawings to help you or someone else to remember this equation.

carbon dioxide   +   water   →   carbohydrate   +   oxygen

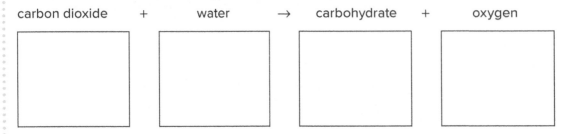

Activity 8: Investigating oxygen production

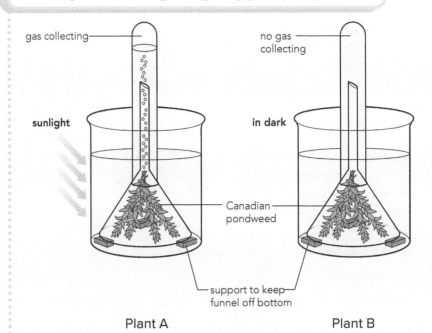

Plant A   Plant B

a) In the investigation shown above, learners put Plant A in sunlight and Plant B in darkness. What were they trying to prove?

....

....

b) What process is taking place in Plant A that is not taking place in Plant B? Why?

....

....

c) What could you do to prove that the gas produced by Plant A is oxygen?

....

....

d) Using the outcomes of the investigation the learners carried out, complete the missing section of the photosynthesis equation below. You will notice that the equation is dual coded; it has words and pictures, as an example of how to memorise equations.

water + carbon dioxide $\longrightarrow$ carbohydrate + oxygen

TAKE A BREAK

When you sit in the same position for a long time, your body can become tense. To relieve some of this tension, take a break from revising and stand up and stretch. This will also help to improve your energy levels and keep you calm. Try bending your neck gently from side to side, stretching out your arms above your head, or standing up and touching your toes without bending your knees.

REMEMBER: MAXIMISING PLANT GROWTH

Knowledge of plant growth, in particular photosynthesis, can be used to create conditions that maximise plant growth. For example, to help farmers grow more and better-quality food, some crops are grown indoors, such as tomatoes, cucumbers and sweet peppers. Farmers also use special lights, which means they do not have to rely on natural sunlight. Using artificial lighting allows plants to photosynthesise 24 hours a day, and also through the winter months when natural sunlight levels are low.

The global population has risen, so more food is needed, and climate change has led to unpredictable weather and growing conditions. Growing plants indoors under artificial lights means that food can be grown all year round and in urban spaces, not just open farmland. This means that people can get more food that is locally grown in a shorter amount of time, and also reduces the distance food travels, which helps to reduce pollution caused by transporting food.

Activity 9: Evaluating a learner's answer

Learners were asked to answer the following question:

Why do commercial plant growers use artificial light throughout the day and at night?

Using the Remember section and your own knowledge about photosynthesis, highlight one strength and one weakness of the model answer below. For the weakness, rewrite the sentence to improve it, so that it would get more marks in a test.

The learner's answer

Commercial plant growers use artificial light so that plants photosynthesise for 24 hours a day. Because plants can photosynthesise 24 hours a day, they can produce more glucose, which is for energy and to make cellulose and starch.

a) Strength ....

....

b) Weakness ....

....

c) Rewrite the answer to improve it.

....

....

....

Activity 10: Marking learners' answers

Below is the mark scheme for the question: *Why do commercial plant growers use waste carbon dioxide from local factories?*

Use the mark scheme to mark each of the following answers out of five. Explain the reason for the marks you have given each answer.

Includes	Mark
Plants need carbon dioxide	1
Equation for photosynthesis	1
Begins answer using the first part of the question	1
Explanation of the role of carbon dioxide in photosynthesis	1
Done so that plants can photosynthesise more and more food is grown	1

Answer A:

They use carbon dioxide because it is needed in photosynthesis.

Mark and reasoning ....

....

Answer B:

Some commercial growers use waste carbon dioxide from local factories because carbon dioxide is needed when plants photosynthesise.

Mark and reasoning ....

....

Answer C:

Some commercial growers use waste carbon dioxide from local factories because carbon dioxide is used by plants, along with light and water, to photosynthesise and create oxygen and glucose. The word equation for this is: carbon dioxide + water + energy from light → glucose + oxygen. This increases food production, grows food more cheaply and uses up waste products.

Mark and reasoning ....

....

Activity 11: Create a test question

Here is an answer from a test paper. Create the question that fits this answer:

Pipes for carrying CO_2 are placed near the leaves because it is the leaves that take in carbon dioxide in the process of photosynthesis.

Question ....

....

Activity 12: Variegated leaves

This is a photograph of a geranium plant. The leaves are variegated, which means that they are different colours, not just green.

Answer the following question, using the mark scheme in the table below to help you achieve all four marks.

How do you think variegated leaves affect the process of photosynthesis?   [4 marks]

Includes	Mark
Use of scientific vocabulary	1
Link made between chlorophyll and colours in variegated leaves	1
Equation for photosynthesis	1
Link made between chlorophyll and photosynthesis	1

....

....

....

Activity 13: Testing for starch

Plant A: Non-variegated geranium   **Plant B: Variegated geranium**

Learners tested the leaves from two geranium plants, A and B, for starch.

Predict what their results should show. Explain the reasons for your predictions.

Plant A ....

....

Plant B ....

....

Revision test

1   Which part of a plant does **not** carry out photosynthesis? Circle the correct answer.   [1 mark]
   A Stem   B Roots   C Leaves

2   Which of these is a product of photosynthesis? Circle the correct answer.   [1 mark]
   A Water   B Minerals
   C Carbon dioxide   D Glucose

3   Write the word equation for photosynthesis. Use these words to help you:
carbohydrate   carbon dioxide   oxygen   sunlight   water   [1 mark]

....

4   Use the correct word equation to explain, using sentences, the process of photosynthesis.   [2 marks]

....

....

....

5   Name two factors in the environment that affect the rate of photosynthesis in a plant.   [2 marks]
   1   ....
   2   ....

6   Describe four stages in testing a leaf for starch and explain each stage. Finally, describe the result of a starch test.   [5 marks]
   1   ....
       ....
   2   ....
       ....
   3   ....
       ....
   4   ....
       ....
   Result: ....

7 If a variegated geranium is kept indoors in low light, the non-green areas of the leaf start to turn green. Explain why this is an advantage for the plant. [2 marks]

....

....

....

....

8 Look at the graph.

a) Suggest a title for this graph. [1 mark]

....

....

b) At what time of the day is photosynthesis at its peak? Suggest why. [2 marks]

....

....

....

c) Why does the rate of photosynthesis slow down during the afternoon? Explain your answer. [2 marks]

....

....

....

d) How much carbon dioxide is absorbed at 2 p.m.? [1 mark]

....

e) If carbon dioxide is being absorbed by the plant, what gas is being produced? [1 mark]

....

Chapter 3 Genetics

REVISION APPROACH: THREE PILES

You are going to begin this chapter using a revision strategy called 'three piles'. The aim of this approach is to help you decide which scientific words you know and can use.

The three piles are shown in the table below.

Know and can use	Not sure	Don't know
I am confident that I know this word and its definition and can use this word correctly.	I have heard of this word; I think I know what it means but I am not sure I am right.	I don't know this word, or I have heard this word but do not know what it means.

The aim as you work through this chapter is to develop your understanding of the key words and move all the words to the 'Know and can use' pile, as you become more confident.

Activity 1: Three piles

a) Make a card for each of the words in the list below. You only need to write the word on the card for now.

adaptation	DNA	genetics
chromosomes	dominant gene	inherit
continuous data	environmental	inherited variation
continuous variation	variation	natural selection
discontinuous data	genes	recessive gene
discontinuous variation	genetic variation	species
	geneticist	variation

b) Look at your word cards and think carefully about each word. Assess how confident you are about the word, then sort the cards into the three piles shown in the table. When you have finished, take a photograph of the piles or copy the three lists into your notebook.

c) Take the word cards from the 'Know and can use' pile and, on the front of each card, add a simple illustration. On the back of each card, write as much information as you can linked to the word. This could include a definition, facts or examples. Keep the cards that you have made in a safe place. You will repeat this activity after each section, so you will be using the cards throughout this chapter.

REMEMBER: SPECIES AND VARIATION

A species is a group of similar organisms that can reproduce (have offspring). Scientists use species to classify (organise into groups) living things. Humans, elephants and cats are all examples of species.

Variation is the difference between individuals within a species. You can see an example of this in the different cats shown in the photograph.

Some variation is passed from parents to offspring, such as hair, eye or skin colour. This is called inherited or genetic variation. Some variation is the result of the surroundings or what an individual does, such as wearing spectacles or having a scar. This is known as environmental variation.

When scientists research variation, they collect two types of data: continuous and discontinuous. Continuous data can have any value, whereas discontinuous data have discrete or distinct values/conditions.

Continuous data	Discontinuous data
Measurement of any value, e.g. in centimetres or grams	Discrete value measurement
Examples: height, mass (weight), hand span	Examples: blood group, eye colour, scars
Data is shown as a line graph	Data is shown as a bar graph

Activity 2: What kind of data?

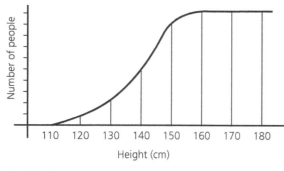

Graph A

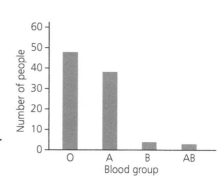

Graph B

Look at Graphs A and B and answer the following questions.

a) Does Graph A show discontinuous or continuous data? Why has this type of graph been used for this data?

....

....

b) Does Graph B show discontinuous or continuous data? Why has this type of graph been used for this data?

....

....

Activity 3: What kind of variation?

Complete the table by deciding whether each example of variation is genetic, environmental or both (tick both columns), and whether data collected about it is continuous or discontinuous.

Variation	Genetic	Environmental	Continuous data	Discontinuous data
skin colour				
height				
foot length				
hair colour				
blood group				
ear piercing				

Activity 4: Check your words

We have now covered the following key words from your word pile:

continuous data   discontinuous data   environmental variation

genetic variation   species   variation

a) Take these words from your word card pile. If you already knew any of these words and created a card for them, check the information to make sure it is correct. Ask someone else to check them as well.

b) For the words that were in the 'Not sure' and 'Don't know' piles, add the information you have learned to each card. On the back of each card, write as much as you can linked to the word. This could include a definition, facts or examples. If it helps you to remember the words, draw pictures, such as a sketch of a bar graph for discontinuous data.

c) Ask someone to test you on each of the words from this section. If you are still unsure of any of the words, find a way of learning them so that by the time you finish this chapter you are confident in using those words.

REMEMBER: CHROMOSOMES AND DNA

All living things are made up of cells. The diagram shows the basic structure of a cell.

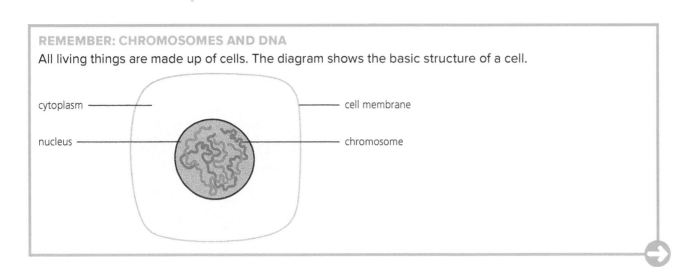

Each cell has a cell membrane, cytoplasm, a nucleus and chromosomes. Chromosomes are thread-like structures located inside the nuclei of animal and plant cells.

Chromosomes are made from a substance called deoxyribonucleic acid, shortened to DNA. Chromosomes contain hundreds or thousands of genes. These genes determine what traits an organism will inherit.

In human reproduction, one copy of each chromosome is inherited from the female parent and one copy from the male parent. This means that children inherit some of their traits from their mother and others from their father.

Different species have different numbers of chromosomes in each cell. Humans have 23 pairs of chromosomes, giving a total of 46 chromosomes in each body cell. A fruit fly has only four pairs of chromosomes in each cell, or eight chromosomes in total. As humans we get 23 chromosomes from our mother and 23 from our father (total 46).

Scientists who research human genetics (called geneticists) number the pairs of chromosomes from 1 to 22. The final pair (23) is known as the 'X/Y' pair (also known as the sex chromosomes). This X/Y pair is important because it controls whether a human is male or female.

The X/Y pair of chromosomes is different for males and females.

- Females have two X chromosomes: XX.
- Males have an X and a Y chromosome: XY.

Activity 5: Size order

Below are four components of a cell:

DNA strand   nucleus   chromosome   gene

Rearrange them in order of size, from largest to smallest.

....

....

Activity 6: Inside a cell

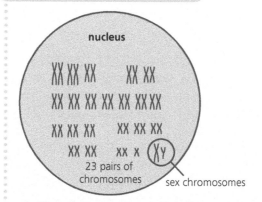

a) The diagram shows the chromosomes in a cell. Is this cell from a male or a female? Explain how you know.

....

....

b) On the diagram below, draw the 23rd pair of chromosomes for a cell of the opposite sex to the cell shown in part a).

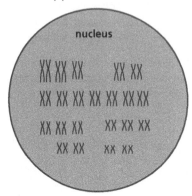

Activity 7: Three piles

Go back to the word cards you made in Activity 1. Look at how you sorted the words and then re-sort them, for example by moving words from the 'Not sure' group to the 'Know and can use' group. Which words are you still unsure about? Think about how you are going to learn these words before you get to the end of this chapter.

Activity 8: Gregor Mendel

A learner decided to make notes on the work of Mendel using short pieces of text and illustrations. Below is the learner's work. However, the learner is unsure about which images to match with each piece of text. Read the learner's work and complete the table to match each piece of text with the correct image. The first one has been done for you.

Text

A   Gregor Johann Mendel (1822–1884) was an Austrian monk/botanist. He studied pea plants. He found that certain features (characteristics) in one generation of pea plants were passed on to the next. These characteristics included: plant height, seed colour, flower colour, wrinkly or smooth seeds.

B   Mendel used knowledge of plant fertilisation to cross-fertilise pea plants with different characteristics to see what the offspring were like.

C   Using his data, Mendel concluded that all living things, including humans, passed on their characteristics to their offspring, and that this could be predicted. For example, he crossed a short plant with a tall plant, and the new plants were all tall. He then harvested these plants for new seeds, but noted that some of the new offspring were short plants. His data showed that there were three times as many tall plants as there were short plants and the ratio was 3:1.

D   Each feature was an inherited factor, each factor had two sets of instructions, and each parent passed on one set of instructions to their offspring.

E Years later, Mendel's characteristics or factors were discovered to be genes.

F Mendel's work has been applied to humans; for example, in studying eye colour.

Images

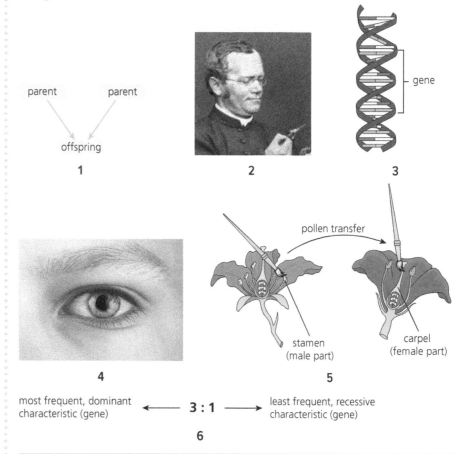

parent   parent

offspring

1   2   3

gene

pollen transfer

stamen (male part)   carpel (female part)

4   5

most frequent, dominant characteristic (gene) ← **3 : 1** → least frequent, recessive characteristic (gene)

6

Text	Image
A	2

REMEMBER: DOMINANT AND RECESSIVE GENES

Geneticists use Mendel's studies to help them work out how characteristics such as eye colour are inherited in humans. Carroll diagrams (also known as punnet squares) can be used to help determine how traits will be inherited.

This Carroll diagram has been produced to show eye colour from two parents.

● Both parents have a dominant (B) and a recessive (b) eye colour gene. This is shown as Bb for both parents.

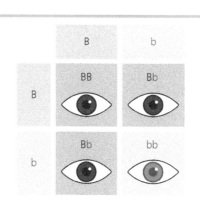

- When the parents reproduce, the likely eye colour of their offspring can be worked out by looking at how the different genes they each provide can combine.
- If the dominant genes from the two parents combine, the child will have BB for their eye colour, meaning they will have brown eyes.
- If the dominant gene from one parent combines with a recessive gene from the other, the dominant gene will be expressed and the eye colour will again be brown.

- If the recessive genes from the two parents combine, the child will have bb for their eye colour, meaning they will have blue eyes.

This eye colour example gives a good overview of how traits are inherited, but it is not the whole story. In reality, there are several different genes that affect eye colour, not just one. This explains why there are so many different possible eye colours (dark brown, hazel, green, light blue, etc.) rather than just two, and why two blue-eyed parents can have a brown-eyed child.

Activity 9: Carrol diagram (or punnet square)

TAKE A BREAK

Chat to a friend for 5 minutes about something completely different from what you are learning, to give your brain a break. Then return to your revision.

For each gene combination, work out what the eye colours would be and draw them into the Carroll diagram.

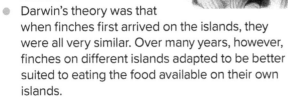

	b	b
B		
b		

REMEMBER: NATURAL SELECTION

- Charles Darwin visited the Galapagos Islands, where he noticed that the same species of animal looked different on different islands. For example, finches had different beaks on different islands.
- Darwin's theory was that when finches first arrived on the islands, they were all very similar. Over many years, however, finches on different islands adapted to be better suited to eating the food available on their own islands.

- Finches that had beaks suited to the food on their island survived and passed these traits on to their offspring. The beaks of each new generation therefore became gradually better suited to the food available. Those finches whose beaks were not suited to the food available died out, so did not pass on their traits to offspring.
- Eventually, all the finches on each island had beaks suited to the food available on their island – we say they had 'evolved'. Darwin called this process natural selection.
- Darwin collected lots of data and evidence about the finches, and used it to support his theory of natural selection.

Activity 10: Changing beaks

Complete this series of four drawings by drawing a sketch in boxes 2 and 3 to show how the beaks of finches evolved over thousands of years, from beaks for eating seeds to beaks for eating insects.

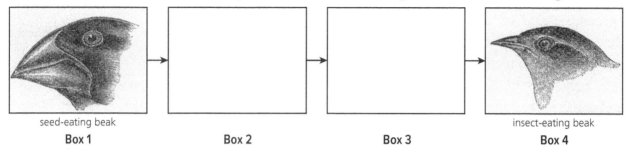

seed-eating beak			insect-eating beak
Box 1	**Box 2**	**Box 3**	**Box 4**

Activity 11: Galapagos tortoise

Another animal that has adapted to its environment in the Galapagos Islands is the tortoise.

Tortoise A   Tortoise B

Tortoise A has a domed shell and short legs. It eats mainly grass and other plants that grow on the ground.

Tortoise B lives on a different island where there is very little ground vegetation, but there are lots of shrubs and small trees. Describe how Tortoise B has evolved to live on that island.

....

....

....

....

....

....

....

....

....

....

Activity 12: Apply your understanding

Learners were asked to answer this question:

Think back to your revision of genes and inheritance. How do genes and inheritance apply to Darwin's theory of adaptation and natural selection in finches on the Galapagos Islands? [6 marks]

The mark scheme for the question is below.

Includes	Mark
There was variation among the finches that landed on the Galapagos Islands.	1
Finches with beaks suited to their environment survived (1 mark) and had offspring, who inherited genes, e.g. for long beaks, from their parents' chromosomes (1 mark).	2
Over time, the most successful adaptations, e.g. long beaks (1 mark), were genetically inherited from generation to generation (1 mark).	2
The birds evolved; those that did not adapt became extinct.	1

Here are two different answers from learners. Use the mark scheme to assess their work.

Answer A:

We can apply the ideas of genes, chromosomes and inheritance to Darwin's theory because the birds passed their long beak genes to their offspring, which inherited long beaks.

Mark and reasoning ....

....

Answer B:

We can apply the ideas of genes and inheritance to Darwin's theory because there was variation amongst the finches. Those birds that had features (for example, long beaks) suited to the environment survived and passed on in the chromosomes the feature (for example, the gene for long beaks) to their young. These birds also passed on that gene to future generations and they formed a new species. Those birds that did not have the feature and could not adapt to the environment died out; they became extinct.

Mark and reasoning ....

....

Activity 13: Scientific ideas

Scientific ideas evolve over time, just like animals and plants, as one scientist's ideas are checked and challenged by later scientists. Read through the information in the table and then draw a line to match each scientist to their idea. If you are unsure, carry out some additional research.

Aristotle	**1** Studied the structure of molecules by firing x-rays at them. Investigated DNA using this method – the results suggested that DNA could be made of two coiled strands.
Mendel	**2** Suggested the idea that plants' features were inherited and that each feature was controlled by an inherited factor. Also suggested that each factor has two sets of instructions and that parents pass on one set of instructions each to their offspring.
Darwin	**3** Worked out that DNA is made from long strands of chemicals that are coiled together to make a structure called a double helix.
Rosalind Franklin	**4** Proposed that animals and plants that have the best features for surviving in their environment have a better chance of surviving than those that do not. This was called natural selection.
Francis Crick and James Watson	**5** Suggested the idea that living things could be grouped and placed on a 'ladder', with the simplest on the bottom rung and the most complex on the top. Thought that species do not change.

Activity 14: Check your words

In this section, the following key words have been used:

adaptation   chromosome   DNA   dominant gene
geneticist   natural selection   recessive gene   species   variation

a) Take these words from your word card pile. If you already knew any of these words and created a card for them, check the information on each card to make sure it is correct. Ask someone else to check those cards as well.

b) For the words that were in the 'Not sure' and 'Don't know' piles, create a card for each word. On the back of each card, write as much information as you can linked to the word. If it helps you to remember the words then you could also draw pictures.

c) Ask someone to test you on each of the words from this section. If you are still unsure of any of the words, find a way of learning them so that you become confident in using those words.

d) Go back over the whole set of cards for the whole chapter. Are there any words that you are still unsure of or do not know? Go back over some of the revision activities to see if they can help you remember. Ask your teacher if there are any words you are still unsure about.

TIPS FOR SUCCESS

Go back over the work that you have done in this chapter to remind yourself of all the information you have covered. Remember to look at revision aids, such as the three piles. When you are ready, complete this short test.

As you work through it, you can help yourself by:

● reading each question carefully – check you understand the question
● looking for key words to use in your answer
● answering the question in your mind first, before you write it down
● making sure you use correct scientific vocabulary in your answers
● using a piece of spare paper to draft any extended answers first, then when you are happy with it you can write your answer in this book
● checking your answers to make sure that you do not want to make any changes.

Revision test

1 What is variation? Circle the correct answer.   [1 mark]
  A Differences between individuals of the same species
  B Differences in where an animal or plant lives
  C Differences in the way organisms eat

2 What is genetic variation? Circle the correct answer.   [1 mark]
  A Variation in the environment
  B Variation passed from parents to offspring
  C Variation in habitats

3 What is the difference between continuous data and discontinuous data?   [1 mark]

....

....

....

4 Sort these data sets and place them in the right places in the Venn diagram.   [8 marks]

blood group   ear piercing   eye colour   hair colour
height   scars   skin colour   tattoos

5 How many chromosomes do human cells have? .... [1 mark]

6 Which substance in a cell stores the genetic information? .... [1 mark]

7 Which sex chromosomes do females have? .... [1 mark]

8 Which sex chromosomes do males have? .... [1 mark]

9 List two things that Mendel contributed to our understanding of genetics and inheritance. [2 marks]

....

....

10 Complete the Carroll diagram (punnet square) on eye colour. [2 marks]

	B	B
b		
b		

11 Explain Darwin's theory of natural selection using your understanding of genetics and inheritance. You must use each of the following words in your answer. [4 marks]

adapted   environment   evolution   extinct   traits offspring   genetic
inheritance genes   species   variation

....

....

....

....

....

....

....

....

Chapter 4 Care in fetal development

CHAPTER INFORMATION

This chapter will help you to revise how a developing baby depends on its mother during pregnancy. By the end of this chapter you should be able to describe:

- how fetal development and the mother's health can be affected by the mother's diet, smoking and drug use
- how diseases can affect fetal development
- how incubators are used in the care of neonates (newborn babies).

REMEMBER: PREGNANCY

Someone who is going to have a baby is said to be pregnant. Pregnancy lasts for about 280 days or 40 weeks in humans, which is about 9 months. During pregnancy, the baby develops inside the mother's uterus, where it is known as a fetus.

The fetus is attached by an umbilical cord to its mother via the placenta, which is attached to the uterus. The umbilical cord is very important. This tube-like cord takes blood containing nutrients and oxygen from the placenta into the fetus, and carries blood containing waste materials out of the fetus.

You could think of the umbilical cord as a lifeline for the developing fetus, so it is important that the mother looks after herself during pregnancy. Whatever the mother eats, drinks, smokes or takes in the form of prescribed or non-prescribed drugs has the potential to affect the fetus. This is because the substances get into the mother's bloodstream and can pass from the mother, through the umbilical cord, to the fetus.

Activity 1: Pregnancy

Label the diagram by filling in the missing labels.

a) _____

b) _____

c) _____

b) _____

Activity 2: Umbilical cord

Some people call the umbilical cord the developing baby's lifeline. Explain why.

....

....

....

....

Activity 3: Health and fetal development

Learners were asked to answer the following question:

Why should a pregnant person be careful of what they eat, drink and smoke during pregnancy?

Here are three answers. Mark each one out of 4 and explain the marks that you awarded.

Answer 1:

Because it can harm the baby.

Mark and reasoning ....

....

Answer 2:

Because the smoke or drink can harm the fetus by going through the umbilical cord to the fetus.

Mark and reasoning ....

....

Answer 3:

What a pregnant person eats, drinks or smokes can affect the fetus. This is because substances they contain can pass into the mother's bloodstream and be carried to the placenta, and so through the umbilical cord into the blood of the fetus.

Mark and reasoning ....

....

REVISION APPROACH: INFOGRAPHICS

An infographic is made using pictures, charts and graphs so that information can be read easily. Using infographics can help you to remember information, as it draws on the idea of 'dual coding'. Dual coding provides two different representations of the information, such as both words and pictures at the same time. In the first section of this chapter, you will use an infographic to prompt your memory and deepen learning about fetal development.

Read through the infographic and discuss it with someone else. Think about what in the infographic you already know and what might be new. Also think about which ideas link together in the infographic.

Alcohol

Drinking alcohol during pregnancy can lead to Fetal Alcohol Syndrome Disorder (FASD):

- low birth weight
- emotional problems
- abnormal eyes, lips, nose
- learning difficulties
- physical disability
- hyperactivity.

Drugs

Drugs prescribed by a doctor, from a pharmacist or non-medicinal (or 'recreational') drugs can all affect the unborn baby.
The liver of a fetus is not able to process drugs. This can lead to:

- premature babies
- underweight babies
- stillborn babies
- behavioural problems in childhood.

Diet

Pregnant women need to eat a balanced diet to receive all the nutrients needed for their own health and the health and growth of the fetus.

Smoking

one cigarette

↓

over 4000 chemicals, many of them toxic

↓

pass through placenta to fetus

↓

stillborn or premature babies, low birth weight, birth defects

Globally, 2.4 million newborns die each year.

Most of those newborns and pregnant women die due to preventable and treatable reasons.

Poverty can prevent people from having a balanced diet. The mother's health may suffer and the growth of the fetus slows down. The baby is then born smaller and less able to fight off disease.

Highest infant mortality rate: Afghanistan

110.6 deaths per 1000 children of 5 years old and younger.

Lowest infant mortality rate: Monaco

1.8 deaths per 1000 children of 5 years old and younger.

Activity 4: 3–2–1

a) When you have read and discussed the infographic with someone, write down three things that you already knew about how fetal development can be affected by what the mother does.

1 ....

2 ....

3 ....

b) Write down two things that you did not know and describe how you will remember them.

1 ....

....

2 ....

....

c) Write down one thing that surprised or shocked you. Explain why.

1 ....

....

Activity 5: Mortality rate

Why do you think there is such a big difference between the infant mortality rates in Afghanistan and Monaco? Explain your reasons. If you are unsure about either country, do some research about them.

....

....

....

Activity 6: Smoking

A pregnant person made this statement about smoking:

'I have been smoking for years – I can't stop ... even though I am pregnant.'

What evidence could you use to convince this pregnant person to stop smoking?

....

....

....

Activity 7: Drugs

Explain how prescribed or recreational drugs can pass from the mother to the fetus and what effect drugs can have on a fetus.

....

....

....

TAKE A BREAK

Take a break to breathe slowly. Doing this helps to relieve stress and tension in your body and can help to get more oxygen into your blood, which helps the brain to work better.

- Position yourself so that you are comfortable but sitting up straight, with shoulders back.
- Close your eyes and focus your attention on your breathing.
- Breathe in for a count of three and out for a count of three. Repeat this several times.

REMEMBER: EFFECTS OF DISEASES ON THE FETUS

Some diseases can have an adverse (harmful) effect on a fetus. Pregnancy can make some women more vulnerable to (at risk of) infections; sometimes even a mild infection can make a pregnant mother very ill.

Any illness the mother gets, such as malaria or tuberculosis, can also cross the placenta to the fetus, which can threaten the fetus' health. So, pregnant women are advised to take extra precautions to avoid becoming ill; for example, being strict about hygiene and washing hands after going to the toilet and before and after handling food. They should also keep away from anyone with an infectious disease.

Sexually transmitted diseases (STDs) can be transmitted from the mother to the fetus during pregnancy; for example, chlamydia, gonorrhoea, syphilis and HIV. These diseases can result in a wide range of problems for newborn babies, such as:

- low birth weight (less than 2 kg (4–4.5 lbs)
- pneumonia
- brain damage
- blindness
- deafness
- hepatitis
- meningitis
- stillbirth.

Activity 8: Diseases commonly affecting fetuses

Find out which parts of the body the following diseases affect.

a) Pneumonia ....

b) Meningitis ....

c) Hepatitis ....

Activity 9: Healthcare communication

List two different ways that healthcare workers could communicate with women to educate them about the dangers to the fetus of a mother catching a disease during pregnancy. Explain why each way could be successful in teaching mothers about those dangers.

1 ....

....

2 ....

....

REMEMBER: NEONATAL CARE AND TECHNOLOGY

A newborn baby is also called a neonate. Neonatal refers to the period from birth to 4 weeks old. This is a very important time because it is during these first weeks that a baby is at the highest risk of dying. Some babies are at a higher risk than others; for example, babies that are premature (born before 37 weeks of pregnancy), or a baby that weighs less than around 2 kg (4–4.5 lbs) at birth. Other risks include those mentioned in the infographic earlier in this chapter.

Neonates that are at risk are given neonatal care in hospital in a neonatal unit. You will notice that these words all have 'neo' and 'natal' in common. The prefix 'neo-' means 'new', and 'natal' means 'relating to birth'. Neonatal care often includes the use of incubators to keep the baby warm, provide it with oxygen and monitor it 24 hours a day. Many incubators include a range of different technologies to monitor neonates in the first weeks of life. Here are some examples.

Temperature sensors placed on the neonate's skin activate an alarm if the neonate's temperature is too low. Incubators keep the baby warm.

If neonates are premature, their lungs might not have developed fully and they may have difficulty breathing. They can be placed on a ventilator to help them breathe.

Air filters keep the air in the incubator clean and free of harmful particles.

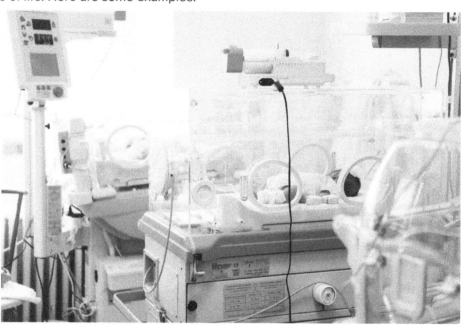

Portholes allow health workers and parents to handle the neonate without passing on germs. The portholes often have rubber gloves built into them.

A vital signs monitor is a machine with small pads that are placed on the baby's chest. It checks that the heart is beating properly and sets off an alarm if there are problems.

An oxygen saturation monitor uses sensors placed on the baby's hand or foot. It checks the amount of oxygen in the neonate's blood and sets off an alarm if the oxygen level is too low.

Activity 10: Ten words or less

Use the information from the Remember section to complete the table below. You should use no more than **ten words** to describe the purpose of each piece of equipment.

Technology	Purpose
air filter	
oxygen saturation monitor	
temperature sensor	
ventilator	
portholes	
vital signs monitor	

Activity 11: Neonate definition

Define the term 'neonate' by describing what each part of the word means.

....

....

....

Activity 12: Check your learning

Go back over the work you have done in this chapter. Work with someone else and ask each other questions about the information in the chapter. Which questions did you find hard to answer? Go back and read that section again, making notes and writing down key words or drawing pictures to help you remember.

TIPS FOR SUCCESS

Go back over the work that you have done in this chapter to remind yourself of all the information you have covered. Remember to look at revision aids, such as the three piles. When you are ready, complete this short test.

As you work through it, you can help yourself by:

- reading each question carefully – check you understand the question
- looking for key words to use in your answer
- answering the question in your mind first, before you write it down
- making sure you use correct scientific vocabulary in your answers
- using a piece of spare paper to draft any extended answers first, then when you are happy with it you can write your answer in this book
- checking your answers to make sure that you do not want to make any changes.

Revision test

1 What is a fetus? Circle the correct answer. [1 mark]
  A A developing baby during pregnancy
  B A newborn baby
  C A developing child over 2 years old

2 What is a function of the umbilical cord? Circle the correct answer. [1 mark]
  A Carries nutrients and oxygen from the fetus to the mother
  B Carries waste material from the placenta to the fetus
  C Carries nutrients and oxygen from the placenta to the fetus

3 Complete the following table by ticking the correct column to show whether each statement is true or false. [3 marks]

Statement	True	False
The average pregnancy lasts 50 weeks.		
Mothers can take drugs during pregnancy without harming the fetus.		
Cigarettes contain chemicals that are harmful to both mother and fetus.		

4 Write definitions for each of the following: [3 marks]

  a) Neonate ....

  b) Umbilical cord ....

  c) Premature baby ....

5 Write four different ways that sexually transmitted diseases (STDs) can affect a newborn baby. [4 marks]

....

....

....

....

6 Explain how chemicals from cigarettes are passed from the mother to the fetus. [3 marks]

....

....

....

7 Explain why a neonate might need to be placed in an incubator. [2 marks]

....

....

8 Name three items of technology that can be found in an incubator and describe how they help to keep a neonate alive. [6 marks]

  1 ....

  2 ....

  3 ....

Chapter 5 Environmental change and extinction

CHAPTER INFORMATION

This chapter will help you to revise your learning about how environmental change and the process of extinction are linked. By the end of this chapter you should be able to describe:

- Jane Goodall's research on chimpanzees
- how environmental changes can affect the populations of organisms
- how a species can become extinct because of environmental change
- how humans can cause environmental change
- global extinction events on Earth.

REVISION APPROACH: HEXAGON MAP

As in previous chapters, you are going to make key word cards for this topic. However, this time the cards will be hexagonal. Hexagons tessellate, which means that you can place your hexagon-shaped cards next to each other and make lots of links. The information on each hexagon must link with the hexagons next to it. Here is an example:

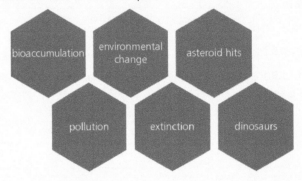

Activity 1: Hexagon key word cards

a) Cut out hexagon-shaped cards. Write each of the following key words on the front of a hexagon. Add any images that will help you remember what the word means.

asteroid	food chain	habitats
bioaccumulation	food web	Jane Goodall
carbon dioxide	fossil	Ordovician
dinosaurs	glaciers	oxygen
endangered species	global warming	Permian
	Great Dying	Red List
extinct	habitat surveys	Triassic

b) On the back of each hexagon card, write a definition. Add as many other ideas as you can that link to the key word on the front. Look back at Activity 1 in Chapter 1 to see what else your cards might include.

c) Do not worry if you do not know all the key words yet. You can add to them as you work through this chapter.

Activity 2: Linking hexagon cards

a) Group the hexagons according to the information on each card. Now arrange the hexagons on a sheet of paper or card so that they fit together. Remember, the information on each hexagon must link to the hexagons around it. This is to help you organise what you know. As you work through this chapter, add any new cards to your arrangement, even if it means moving some of the cards you have already put down.

b) Take a photograph to record how you have laid out your hexagons. Keep this photograph and return to it later, to see if you want to move any cards to make new/stronger connections between the hexagons.

c) Work with someone else. Choose two or three of their hexagons and ask them to explain in detail why they have placed them next to each other. What are the links between them?

d) Swap over so that they can ask you the same questions.

REMEMBER: JANE GOODALL

Jane studied chimpanzees in Tanzania for 60 years. She used an unusual approach to studying them: rather than just observing, she lived among the same group of chimpanzees for years and observed how they lived and interacted with each other.

- Jane kept field notes in a journal, recording her observations day by day for years.
- She gave the chimpanzees names, recorded how they could be recognised and noted how they interacted with each other.
- She imitated (copied) what the chimpanzees did, and they became used to her and interacted with her.

Before Jane's research, scientists thought that only humans interacted socially and used tools; they also believed that chimpanzees were herbivores. By recording the daily lives of chimpanzees over many years, Jane could prove that her observations were not one-off events, but happened regularly. She proved that:

- chimpanzees were omnivores, not herbivores; they ate fruits and berries but also hunted and ate small animals
- chimpanzees made and used tools; for example, to get ants out of nests to eat
- chimpanzees interacted with one another socially and emotionally; for example, chimpanzees comfort a mother if her baby dies.

As Jane worked with chimpanzees, she realised that their habitats, and therefore their lives, were being threatened. She saw that the changes that were taking place were devastating chimpanzee habitats; for example, mining, logging and removing trees for agriculture. So she set up the Jane Goodall Institute, which aims to protect chimpanzees and their habitats by working with local people.

Activity 3: 5–4–3–2–1 Book blurb

Imagine that you have just written a biography about Jane Goodall. The publisher has asked you to write the blurb (which should be a maximum of 150 words) to tell readers what the book is about. They have asked that you include:

5 key things Jane found out

4 ways she ensured her research findings were trustworthy

3 ways she changed people's understanding about chimpanzees

2 ways we can help support chimpanzees

1 change everyone can make to conserve local environments.

Write your blurb on a separate piece of paper and ask someone else to check it.

REMEMBER: STUDYING HABITATS

Learners in Kenya have been studying the work of Jane Goodall. They were inspired to set up a group to study the plants and animals in a small area of their school. The learners:

- made notes about habitats in the area
- observed which plants and animals were found in each habitat, took photographs and created tally charts of numbers of animals
- repeated observations daily for three months at regular times: 9 a.m., 1 p.m. and 3 p.m.
- noted key timings in the school day that could affect the results, such as 12 noon, when learners ate their lunch outdoors, and 3 p.m., when the school day ended
- observed how animals behaved
- used quadrats to estimate the population of plants in the school field, then repeated this over three months to monitor changes in the plant population and plant life cycles.

REVISION APPROACH: DOUBLE BUBBLE

The double bubble helps you look at two different ideas, objects or events and compare them to find similarities and differences. The double bubble is a way of helping you to organise and think more deeply about what you have learned. You are going to use a double bubble to compare how the learners from the Kenyan school studied plants and animals in their school grounds with how Jane Goodall researched chimpanzees.

Activity 4: Double bubble

Copy out the double bubble template below on a separate piece of paper. Make sure you make each part big enough to write in.

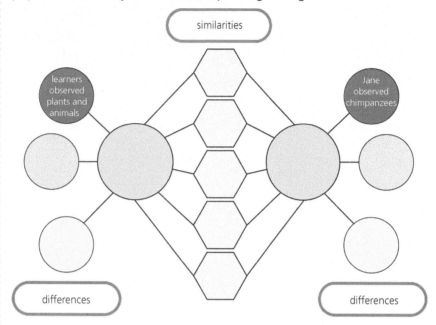

- Start by putting the two ideas you are comparing in the large circles. In the left-hand large bubble, write 'learners'. In the right-hand large bubble, write 'Jane Goodall'.
- Next, put the similarities between the two ideas in the hexagons down the middle.
- Put the differences between the two ideas in the outer coloured circles, making sure differences in the same-coloured circles link across both sides of the diagram.

Activity 5: Using data from quadrats

The learners at the school in Kenya used a 1 m² quadrat and carried out a survey of plants and animals in the school field.

The learners measured the area and decided that it would take 100 quadrats to cover the whole field. To save time, they used a 1 m² quadrat and decided that they would estimate the population for the whole area. Table 1 shows some data from the 1 m² quadrat, collected at 3 p.m. one day.

a) The learners started to calculate the estimated number of animals and plants for the whole field. How did the learners calculate the figures to fill in the right-hand column of the table?

b) Using the same mathematical approach, complete the table.

Table 1: Data from quadrat survey at 3 p.m.

Plant or animal and their food (if animal)	Number in 1 m² quadrat	Estimated number in 100 m² area
grass clumps	18	1800
worms (eat dead plant material)	2	
rove beetles (eat beetle larvae, aphids and small caterpillars)	7	
dandelions	6	
ants (attracted to sweet and sugary foods)	100	
caterpillars (eat leaves)	1	

Activity 6: Using quadrat data

a) Look at Table 1 in Activity 5. The learners had counted the ants in the quadrat at 9 a.m. that day and only found 15. The learners ate their lunch outside on the day the data was taken. Suggest why they found so many more ants in their quadrat at 3 p.m. than at 9 a.m.

....

....

....

....

b) Look at the terms below. What term best describes what kind of enquiry this is? Give a reason for your answer.

observation over time   fair test
research using internet   pattern-seeking

....

....

c) What do you think were the limitations in the way the learners gathered their data? What could they have done differently to increase the validity of their data?

....

....

....

....

....

....

Activity 7: Bird observations and data

Here is the data that learners collected about the number of birds and their feeding habits, which they observed in the environment during their investigation.

Table 2: Bird observation data

Bird	Food	Number of individuals seen at:		
		9 a.m.	1 p.m.	3 p.m.
sparrow	small bread scraps, grains and seeds	3	14	10
speckled pigeon	seeds, grains and peanut crops	2	2	2
starling	seeds, nuts, berries, grains, bread scraps and invertebrates such as spiders, larvae and worms	5	28	16
common bulbul	fruit, nectar, seeds and insects	6	3	4
speckled mousebird	fruits, berries, leaves, buds, flowers, nectar and seeds	2	1	3
black kite	fish, small mammals, birds, bats and rodents	0	1	1

What is the pattern in the data for sparrows and starlings? Explain the reasons for this pattern. Use information from Activity 6 and Table 2 in your answer. Think about what type of food each bird species eats and what the pupils at the school were doing at different times.

....

....

....

Activity 8: Relationship between plants and animals

a) Use the data from Table 1 and Table 2 to create a food chain in the space provided.

b) Which birds would be most affected if ants and beetles disappeared? Explain why.

....

....

c) The school would like to create a database of observations to find out what happens to populations of plants, birds and insects over long periods of time. What advice would you give to the learners in Kenya to make sure that their evidence is reliable?

....

....

....

Activity 9: Revisit hexagons

Go back to the last photo you took of how you laid out your hexagons. Create some more hexagon cards using your key words from this section, and create a new hexagon map by laying out your cards again to help you make new and more connections. How many more connections can you make?

When you have finished, take a new photograph of your hexagons. Keep this somewhere safe; you will use it at the end of this chapter.

REMEMBER: MASS EXTINCTION EVENTS

Scientists have identified five mass extinction events in the history of the Earth. An extinction event occurs when large numbers of species become extinct in the same period of time. Scientists know about these events through studying fossils, and suggest some of the possible causes of what happened. However, as humans did not witness these events and the evidence is millions of years old, not all scientists agree with each other and they have different theories. For example, some scientists believe that dinosaurs became extinct due to catastrophic volcanic eruptions, while others think an asteroid hit the Earth, causing changes in the environment that led to their extinction.

Scientists do agree that during these mass extinctions there were extreme changes to the environment, and these changes were so severe that many animals became extinct. Today, scientists think that the changes to the Earth caused by human activity will lead to another mass extinction.

The following table gives information about each event. Use the table to complete the questions in the rest of this section.

Table 3: Mass extinction events

Extinction event	What period? (millions of years ago)	Possible causes and effects
Ordovician	455–430	Ice age, glaciers, volcanic activity 85% of species extinct, including Brachiopods, trilobites
Late Devonian	376–360	Rapid sea level rise and fall, reduction of oxygen and carbon dioxide 75% of species extinct, including corals, placoderms
Permian/Triassic	252	Volcanic eruptions, asteroid impact, atmosphere filled with dust, Sun blocked out 95% marine invertebrates and 70% land invertebrates died – known as the Great Dying Took 10 million years to recover
Triassic–Jurassic	201	Climate change, possible asteroid impact or volcanic eruptions 76% of species extinct
Cretaceous–Tertiary	66	Asteroid 75% of species extinct, including dinosaurs, pterosaurs, ammonites

Activity 10: Timeline

Below is a timeline of the major extinction events on Earth, but it is not complete. You will need to put the key periods in the correct places on the timeline. Then provide information in each box about the period in which a major extinction event took place. You should only put the following four facts in each box:

- name of period
- number of years ago
- key extinction events
- percentage of species extinct.

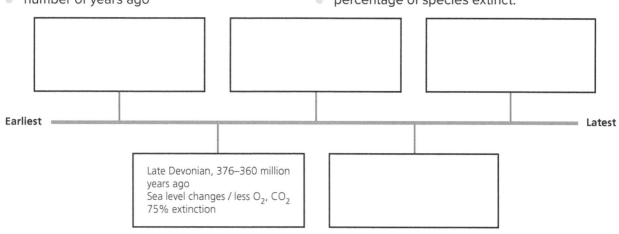

Earliest   **Latest**

Late Devonian, 376–360 million years ago
Sea level changes / less O_2, CO_2
75% extinction

REMEMBER: ANTHROPOCENE EPOCH

Geologists have named the current period of Earth's history the Anthropocene Epoch. In this period, human activity has had a direct impact on the planet's climate and environments. Human activity is likely to cause many species of plants and animals to become extinct much faster.

Our use of fossil fuels, deforestation, agriculture and industry have all dramatically changed the environment and caused a huge increase in the amount of carbon dioxide in the atmosphere. This is resulting in climate change. The Earth is getting warmer, causing glaciers to melt, sea levels to rise and more extreme weather events such as droughts and hurricanes.

These environmental changes are happening so fast that plants and animals are unable to adapt and many are facing extinction. Some scientists think that the Earth is heading towards another mass extinction: the Anthropocene Extinction. As humans are causing these changes, it is humans who must find solutions.

Activity 11: Next mass extinction event?

Suggest why some scientists believe that the Earth is heading for another mass extinction event. Give your reasons.

....

....

....

....

Activity 12: Recently extinct animals

a) Below is a list of animals that have become extinct in the past hundred years. Choose two animals to research and explain why they became extinct.

Pinta giant tortoise   Paradise parrot
West African black rhinoceros   Guam flying fox
Golden toad   Hawai'i 'ō'ō

1 Animal and reason for extinction

....

....

2 Animal and reason for extinction

....

....

b) Look back at the animals you researched. List any common causes of their extinction.

....

....

REMEMBER: THE IUCN RED LIST OF THREATENED SPECIES

The International Union for Conservation of Nature (IUCN) Red List of Threatened Species is a record of all the threatened plant and animal species across the globe. Below are four animals from the list.

Basking shark

Orangutan

Hawksbill turtle

Dugong

Activity 13: Red List postcard

Choose one of the four animals on the Red List and research why it is endangered. Then use the information that you have found to write a postcard to the government of the country where the animal is from, explaining why the animal is endangered and what could be done to save it from extinction. Use the postcard template below.

Activity 14: Revisit hexagons

Go back to the last photo you took of your hexagons.

a) Create a new hexagon map and include the new words you have learned. Think about how best to connect the cards. For example, you could put the cards for 'mass extinction' and 'Anthropocene Epoch' next to each other, to link human activity with changes in environments and the extinction of animals and plants.

b) To challenge yourself you can:
- shuffle the cards and stack them up. Take cards from the top of the pile in the order they are stacked and place them as a hexagon map, talking about the links to someone else every time you place a card down.
- pick out seven cards. Place one in the centre and the other six cards around it. Explain to someone else the links between the touching cards.

You can take photos of your new hexagon maps if it helps you remember the connections you have made.

TIPS FOR SUCCESS

Go back over the work that you have done in this chapter, including the hexagons and double bubble, to remind yourself of all the information you have covered. When you are ready, complete this short test.

As you work through it, you can help yourself by:

- reading each question carefully – check you understand the question
- looking for key words to use in your answer
- answering the question in your mind first, before you write it down
- making sure you use correct scientific vocabulary in your answers
- using a piece of spare paper to draft any extended answers first, then when you are happy with it you can write your answer in this book
- checking your answers to make sure that you do not want to make any changes.

Revision test

1   Which species did Jane Goodall research? Circle the correct answer.   [1 mark]
    A   Orangutans   B   Possums   C   Chimpanzees   D   Elephants

2   What did Jane Goodall observe this species doing that scientists at the time
    thought only humans did? Circle the correct answers.   [2 marks]
    A   They made and used tools.
    B   They ate only plants.
    C   They interacted with each other, such as comforting others.
    D   They lived in trees.

3   Complete the following table by ticking the correct column to show whether
    each statement is true or false.   [5 marks]

Statement	True	False
Jane Goodall worked with chimpanzees for 60 years.		
Jane Goodall studied chimpanzees in Australia.		
Chimpanzees use tools to get food.		
Chimpanzees are omnivores.		
Goodall set up an institute to work with local people to protect chimpanzees.		

4   Write definitions for each of the following: [3 marks]

    a) The Great Dying ....

        ....

    b) The Red List of Threatened Species ....

        ....

    c) An extinction event ....

        ....

5   Give one thing that most extinction events had in common.   [1 mark]

    ....

Use the graph below to help you answer the questions in this section.

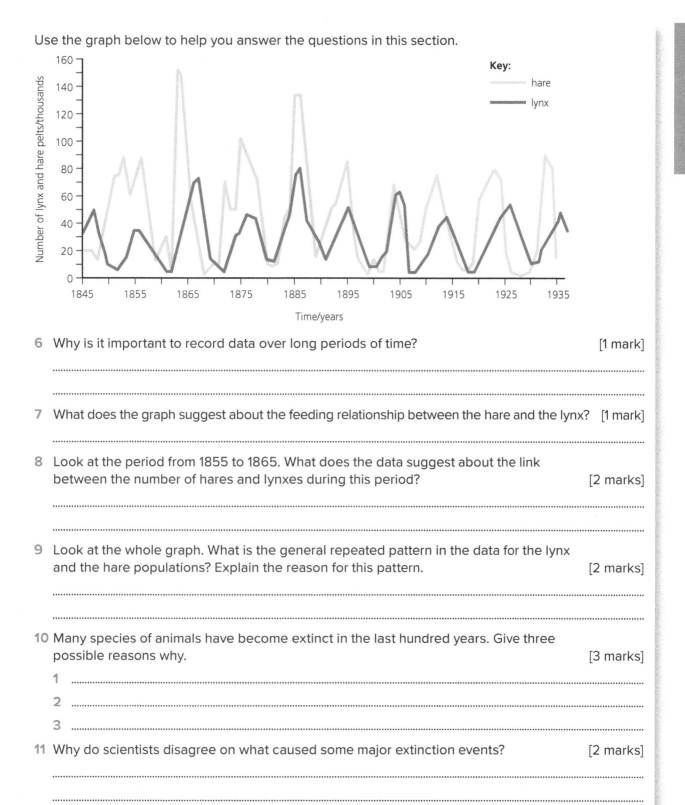

6   Why is it important to record data over long periods of time?   [1 mark]

....

....

7   What does the graph suggest about the feeding relationship between the hare and the lynx?   [1 mark]

....

8   Look at the period from 1855 to 1865. What does the data suggest about the link
    between the number of hares and lynxes during this period?   [2 marks]

....

....

9   Look at the whole graph. What is the general repeated pattern in the data for the lynx
    and the hare populations? Explain the reason for this pattern.   [2 marks]

....

....

10 Many species of animals have become extinct in the last hundred years. Give three
    possible reasons why.   [3 marks]

    1   ....

    2   ....

    3   ....

11 Why do scientists disagree on what caused some major extinction events?   [2 marks]

....

....

....

12 What is the Anthropocene Epoch? Explain some of the main reasons for plant and
    animal species becoming extinct during this epoch.   [3 marks]

....

....

....

Chapter 6 The periodic table

Look at the inside back cover of this book to see a copy of the full periodic table – you will need it throughout this chapter.

REVISION APPROACH: PERIODIC TABLE KEY WORD CONCEPT MAP

In this chapter, you will use a concept map revision strategy, like the one you used in Chapter 1. Concept maps help you to remember and make connections between ideas. They enable you to see what you know and where there are gaps in your learning. You are going to begin by making key word cards to use when you create your concept map.

Activity 1: Key word cards

a) Highlight the words you already know in the list below.

atom	element	noble gas
chemical symbol	group	nucleus
Dalton	Mendeleev	period
electron	negative charge	periodic table
electrostatic attraction	neutron	positive charge

b) Make key word cards for all the words you highlighted. Write each word on the front of a piece of card and add any images that help you remember what the word means.

c) On the back of each card, write as much information as you can linked to the word, such as a definition, facts or examples. Do not worry if you do not know all the words yet; you will repeat this activity at the end of this chapter when you know more words.

Activity 2: Periodic table concept map

a) Sort the cards you know into groups: for example, all about atoms.

b) Lay out the cards you have made on a large piece of plain paper.

c) Draw lines to join as many of the cards as you can on the sheet. On the connecting lines, write the reason why you have linked the cards.

d) The more lines you make, the more links there are that show your knowledge and understanding.

e) When you have finished, take a photograph of your concept map. Keep your concept map word cards safe. You will need them again at the end of this chapter.

REMEMBER: THE PERIODIC TABLE

The periodic table is used by scientists globally; it is a table of all the known elements. Everything in the universe is made up of elements.

An element is a substance that cannot be broken down further by a chemical reaction. There are more than 100 known elements. Every element has a unique atom.

Elements in the periodic table are organised in order of atomic number. The atomic number is based on the number of protons in an atom; the number of protons indicates the identity (name) of an element. The periodic table is organised in a special way (use the periodic table in the back cover of this book to remind you). It has:

- seven rows, known as periods
- 18 columns, known as groups. Some groups have names, such as the alkali metals (Group 1) or the noble gases (Group 0)
- elements grouped into metals and non-metals
- elements placed in order of increasing atomic number
- elements in the same group that have similar properties to each other.

One of the elements, hydrogen, is not placed on the table itself. This is because scientists do not know where to place it. Hydrogen is a non-metal, but it has a structure like Group 1 metals. Scientists call hydrogen an anomaly because it is different and not easily classified.

Activity 3: Properties of elements

a) Draw lines to match each element below with its correct symbol, atomic number and properties.

Element	Symbol	Atomic number	Properties at room temperature
hydrogen	Hg	79	solid, non-metal
carbon	Au	6	liquid, metal
iron	H	80	solid, metal
gold	Fe	1	solid, metal
mercury	C	26	gas, non-metal

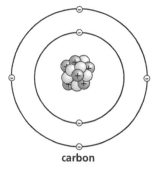

carbon

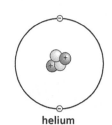

helium

⊕ proton
◯ neutron
⊖ electron

b) Look at the two atoms in the diagram: carbon and helium. State one similarity and one difference between them.

Similarity ....

....

Difference ....

....

REMEMBER: HISTORY OF THE PERIODIC TABLE

The periodic table evolved over several centuries, as understanding about elements developed and scientists built on each other's work.

Activity 4: Periodic table timeline

Read the statements below. Place the correct letter in each timeline box to show how the periodic table was developed over the centuries.

A International year of the periodic table, celebrating 150 years.
B Robert Boyle believed that all substances were made up of atoms, which he called corpuscles.
C John Dalton worked out atomic masses of elements.
D Mendeleev produced his periodic table, which formed the basis of the modern periodic table.
E Lavoisier called the substances that could not be broken down further 'simples'.
F Niels Bohr, working in Copenhagen, published the first explanation of why certain elements fall into particular groups in the periodic table.

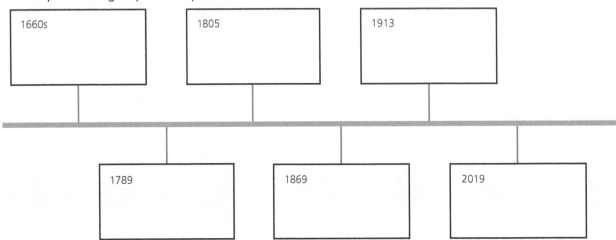

| 1660s | 1805 | 1913 |

| 1789 | 1869 | 2019 |

REMEMBER: MENDELEEV

So you want to know about me! I was born 8 February 1834 in Siberia in Russia. I was one of 14 children – imagine that! I loved chemistry and worked as a teacher and an academic. But I was fascinated by the elements and felt sure that there was a way they could be organised that would help chemists. I wrote every element on a separate card and kept moving them about, a bit like moving playing cards around. When I was ordering them according to atomic weight, I noticed a pattern, with similar properties appearing at regular intervals. I then created the periodic table by arranging chemical elements by atomic mass. I wonder if it is still used?

I led an interesting life – as well as my chemistry work, I helped to design and construct the world's first Arctic icebreaker ship. I even got the Russian government to use the metric system. Oh, and I married twice and had six children. Do you know that I was nominated for the Nobel Prize three times but never won it! But I do have an element named after me – see if you can work out which one it is.

Activity 5: Bookmarking Mendeleev 3, 2, 1

Create a research bookmark about Mendeleev on a separate piece of paper, to help you remember his role in developing the periodic table. You will need to carry out research and think about which information should be included. On your bookmark you should include:

3 facts about Mendeleev's life

2 other things that Mendeleev achieved

1 important feature of his periodic table.

Activity 6: Comparing elements

Group 1 – Alkali metals

Some properties of the Group 1 elements (also known as the alkali metals) are shown below. The graph shows their different melting points.

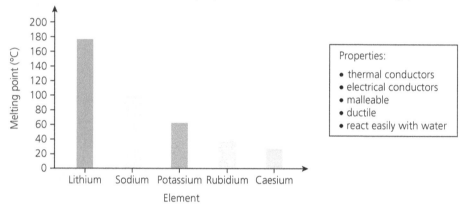

Group 0 – Noble gases

Some properties of the Group 0 elements (also known as the noble gases) are shown below. The graph shows their different boiling points.

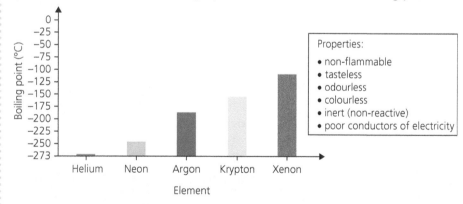

Look at the graphs above and at the periodic table in the back cover of this book and answer the following questions.

a) What is the relationship (link) between the melting points of the Group 1 elements and each Group 1 element's position in the periodic table?

....

b) What is the relationship between the boiling points of the Group 0 elements and each Group 0 element's position in the periodic table?

....

c) Noble gases are also known as monoatomic gases. Find out what the prefix 'mono-' means and then use this to describe what a monoatomic gas is.

....

....

d) Explain the difference between ductile and malleable. Why are these properties important when using metals?

....

....

e) Research how the three alkali metals and three noble gases in the table are used in everyday life. Complete the table below.

Alkali metals		Noble gases	
Name	Uses	Name	Uses
lithium		helium	
sodium		neon	
caesium		radon	

Activity 7: Mnemonics

Look at the first group in the periodic table. Create a mnemonic on a separate piece of paper using the first letter of each element name to help you remember the order of the elements in Group 1.

Share your mnemonic with someone else and tell them which elements the mnemonic is referring to.

Activity 8: Revisit your concept map

Look back at the key words concept map you started in Activity 2.

a) Revisit the key words list in Activity 1 and create new cards for any words you missed.
b) Lay out the concept map word cards again and sort them into groups. Decide on titles for the groups you have made.
c) Place the groups of word cards down on the plain paper. When you are happy with the arrangement, you can glue your cards down and write your titles for the groups above them.
d) Now draw lines to join as many of the cards as you can on the sheet.
e) You can join cards within a group and across different groups. The more lines you make, the more links you can show. Compare this concept map with the one you made at the start of the chapter. How many new connections can you now make? Talk to your teacher about any words you are still unsure of.

TIPS FOR SUCCESS

Go back over the work you have done in this chapter to remind yourself of all the information you have covered. When you are ready, complete this short test.

As you work through it, you can help yourself by:

● reading each question carefully – check you understand the question
● looking for key words to use in your answer
● answering the question in your mind first, before you write it down
● making sure you use correct scientific vocabulary in your answers
● using a piece of spare paper to draft any extended answers first, then when you are happy with it you can write your answer in this book
● checking your answers to make sure that you do not want to make any changes.

Revision test

1 Which is the odd one out? Circle the correct answer.   [1 mark]
   A Dmitri Mendeleev   B John Dalton
   C Albert Einstein   D Robert Boyle

2 Elements in the periodic table are organised in order of atomic number. Which particles in the atom is the atomic number based on? Circle the correct answer.   [1 mark]
   A Protons
   B Neutrons
   C Atoms

3 Which of the following are found in the nucleus of an atom? Circle the correct answer.   [1 mark]
   A Only electrons   B Neutrons and electrons
   C Protons and neutrons   D Only protons

4 Complete the following table by ticking the correct column to show whether each statement is true or false.   [4 marks]

Statement	True	False
No scientists contributed to the periodic table before Mendeleev.		
The periodic table has seven rows and each row is known as a period.		
Elements are placed in the periodic table in order of decreasing atomic number.		
Elements in the same group have similar properties to each other.		

5 Name three properties of alkali metals.   [3 marks]
   1 ....
   2 ....
   3 ....

6 Name three properties of noble gases.   [3 marks]
   1 ....
   2 ....
   3 ....

7 Describe what a monoatomic gas is.   [1 mark]
   ....
   ....

8 Two learners answered the following question:

Explain the difference between the terms 'ductile' and 'malleable'.

Use the mark scheme to **highlight** successful elements of each answer and award marks out of 5. [2 marks]

Includes	Mark
Description of properties of malleable materials	1
Examples of materials that are malleable	1
Description of properties of ductile materials	1
Examples of materials that are ductile	1
Whether materials can have both properties	1

Answer A:
A malleable material is one where the material can be easily beaten with a hammer; copper is an example. Ductile materials are different because they can be stretched into thin wires; for example, gold and copper.

Mark ....

Answer B:
A ductile material is one that can be stretched into thin wires; this is called tensile stress. Gold, copper and silver are ductile materials. A malleable material is one that can be beaten flat with a hammer. Some materials can be malleable and ductile; for example, gold and silver. They can be hammered and stretched into thin wires.

Mark ....

9 Here is part of a learner's concept map. There are five mistakes. Identify them and explain why each is incorrect. [5 marks]

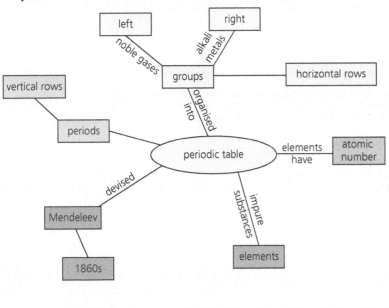

....

....

....

....

....

Chapter 7 Bonds and structures

REVISION APPROACH: BONDS AND STRUCTURES CONCEPT MAP

A concept map is a useful way of helping you remember key words and ideas in a topic. It can also help you check that you have understood how ideas link together.

Activity 1: Revision concept map

a) Look at the list of words below.

anion	electron cloud	neutrons
atom	giant structure	nucleus
bond	ion	orbit
cation	ionic	positive
covalent	molecule	protons
electron	negative	

b) Make word cards for all the words that you recognise – look back to Chapter 1 for an example of a key word card if you need to. Do not worry if you do not know all the words yet. You can make cards for any words you do not know when you meet them as you revise this chapter. Write each word on the front of a piece of card and add any images that help you remember what the word means.

c) On the back of each card, write as much information as you can linked to the word. This could include a definition, facts or examples.

d) Try to sort the cards you know into groups; for example, names of different types of bonding.

e) Lay out the concept map word cards you have made on a sheet of plain paper.

f) Now draw lines on the sheet to connect as many cards as you can. On the connecting lines, write the reason why you have linked the cards. There are lots of different ways that you can link the word cards; you just need to be able to explain your reasoning for the links you make.

g) When you have finished, take a photograph of your concept map and collect and keep your word cards safe. You will need them again at the end of this chapter.

REMEMBER: STRUCTURE OF THE ATOM

You might recall that scientists use a number of models of the structure of the atom. Each model shows a central nucleus, which contains protons and neutrons. They all show electrons around the nucleus, although they show them in different ways. The models for the element beryllium are shown below.

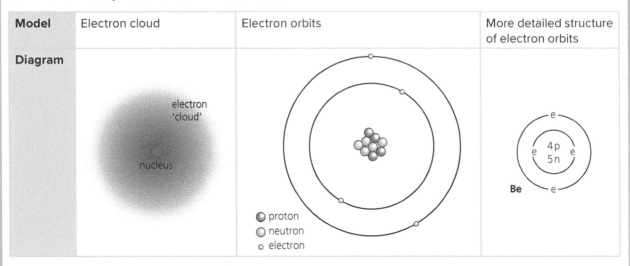

Model	Electron cloud	Electron orbits	More detailed structure of electron orbits
Diagram	electron 'cloud' / nucleus	proton / neutron / electron	Be 4p 5n

Activity 2: Structure of the atom

Some learners were drawing models of an atom of helium, which are shown below.

Helium is a noble gas and has two protons, two neutrons and two electrons in each atom.

Two of the models are correct. One is incorrect.

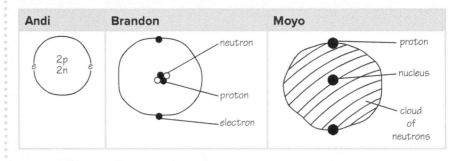

Andi	Brandon	Moyo
e 2p 2n e	neutron / proton / electron	proton / nucleus / cloud of neutrons

Identify the incorrect model and give **two** reasons why it is incorrect.

....

....

....

....

REMEMBER: STABLE ATOMS

In an atom, electrons orbit the nucleus. They fill up the nearest orbit (or shell) to the nucleus first. Each shell has a maximum number of electrons that it can take before it is full.

The atoms of noble gases have full outer shells and so are stable as single atoms. This means that they do not take part in chemical reactions. The noble gases are all in the group in the far right column of the periodic table.

Outer shell	first	second	third
Maximum number of electrons allowed in the shell	2	8	8
Example of atom with full outer shell	helium 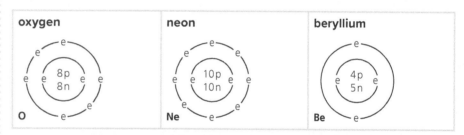 He	neon Ne	argon Ar

If an atom does not have a full outer shell, it is unstable as a single atom. The atom will try to lose, gain or share electrons until it has a full outer shell.

It does this through chemical reactions, which can form covalent or ionic bonds.

Activity 3: Stable or unstable?

Some learners are trying to decide whether some elements are stable as a single atom. Look at the diagrams of the atoms and decide which are stable and which are not.

oxygen	neon	beryllium
8p 8n O	10p 10n Ne	4p 5n Be

Which of these elements are stable as single atoms?

....

REMEMBER: COVALENT BONDING

To become more stable, atoms bond with other atoms. One way of doing this is by covalent bonding. This involves two atoms moving very close so that they can share the electrons in their outer shell, forming a molecule. If the atoms each have seven electrons in their outer shell, the two atoms in a molecule each share one electron. These two electrons form a pair known as a covalent bond.

In some atoms, such as oxygen, the outer shell contains six electrons. This means that to form a covalent bond, each oxygen atom shares two electrons. This is called a double covalent bond, as it involves sharing two pairs of electrons, not just one pair.

Activity 4: Model building

Leon and Jacob are building models of molecules with marshmallows and short strands of spaghetti.

They use marshmallows to represent the atoms and spaghetti to represent the bonds. Each strand of spaghetti represents one bond; double bonds are represented with two strands of spaghetti between the marshmallows.

Leon builds the following covalently bonded molecules:

hydrogen	water	methane	oxygen

Jacob says that almost all the models are correct, except the one for oxygen.

What needs to be done to correct the model for oxygen? Explain why.

....

....

Activity 5: CPK colouring of models

A teacher is using a model to help their learners understand CPK colouring, developed by Robert Corey, Linus Pauling and Walter Koltun. They make some models using sweets.

a) Write in the correct names and the correct structures of the molecules using the correct symbols. One has been completed for you.

Clue: Methane is a hydrocarbon. It only contains atoms of two elements, hydrogen and carbon.

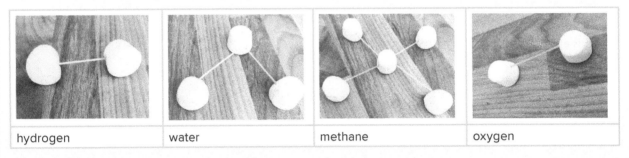

Picture	two white sweets	two red sweets	one black sweet surrounded by four white sweets	one red sweet and two white sweets	two green sweets
Molecule name					chlorine
Structure of molecule				H—O—H	

b) One of the learners says that this model is not very good in terms of showing the bonding of the molecules. Do you agree with them? Give a reason for your answer.

....

....

REMEMBER: IONIC BONDING

Another way that atoms can become stable is by losing or gaining outer electrons. Electrons lost from one atom are usually gained by another atom of a different element in a chemical reaction.

Atoms usually do whatever is easiest to form a full outer shell of electrons. For example, magnesium has two electrons in its outer shell; to get a full shell, it could either lose two electrons or gain six. Since losing two electrons is easier than gaining six, it will lose two electrons.

Electrons are negatively charged. This means that if an atom gains extra electrons, it becomes a negatively charged ion (an anion). If it loses electrons, it becomes a positively charged ion (a cation). As magnesium loses two electrons, it becomes a cation with a charge of +2.

Ions of opposite charge will attract one another and form an ionic bond.

TAKE A BREAK

Stop work and get yourself a drink of water. This will help you keep hydrated, which can help with concentration.

Activity 6: Positive, negative or neither

Use the information in the table to decide if the following atoms will lose or gain electrons to become stable (or do neither). Remember that they will do the simplest thing to gain a full outer shell.

Decide whether the result will be a positive ion, a negative ion or no change.

Add the correct term: **cation**, **anion** or **not an ion**.

Atom	oxygen	neon	magnesium
Diagram	O (8p 8n)	Ne (10p 10n)	Mg (12p 12n)
Will it become a positive ion, negative ion or 'no change'?			
Is it a cation, an anion, or not an ion?			

REMEMBER: SIMPLE MOLECULES

Molecules can be very small or very large. Simple molecules are smaller molecules, perhaps with as few as two atoms in them. Examples include oxygen (two atoms), nitrogen (two atoms) and water (three atoms).

Although the forces between the atoms holding the simple molecule together are very strong, the forces between individual molecules are weak and break easily. This means that substances made from simple molecules are usually liquids and gases at room temperature, as they have very low melting and boiling points. They also do not conduct electricity, as all of the electrons are involved in bonding so they are not free to move around.

Activity 7: Identifying simple molecules

The table below lists the properties of different substances. Use what you know about simple molecules and the properties of these substances to help you decide whether each material is made of simple molecules or not.

Substance	A	B	C
Appearance at room temperature	pale green gas	colourless liquid	colourless crystal solid
Does it conduct electricity?	no	no	no
Is this a simple molecule? (Circle the correct answer.)	yes / no	yes / no	yes / no

REMEMBER: GIANT STRUCTURES

There are two types of giant atomic structures: giant covalent structures and giant ionic structures. Although these large structures are quite similar, there are key differences between them.

You might find information in the textbook about this (pages 79–82), or you can carry out some research of your own.

To help you remember more about these structures, Activity 8 will ask you to complete two Frayer maps, one for each type of structure. The activity contains some questions and clues to help you complete the maps.

Activity 8: Giant atomic structures – Frayer maps

Using the template below, create two separate Frayer maps. Create one for giant covalent structures and one for giant ionic structures. In each map, give a definition, some characteristics, features and facts, examples and non-examples of each type of giant structure.

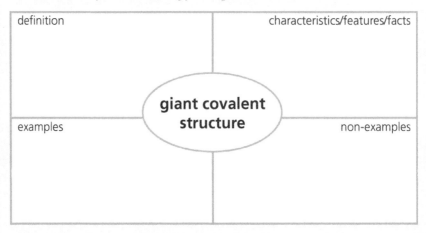

Some key points to consider in each map are:

- an example material
- the properties of the material
- the size of the structure
- the type of bonding
- the shape of the structures
- the strength of the bonding (remember that giant structures have strong bonds between all of the atoms, whether covalent or ionic)
- linking the strength of the bonding to the properties of the material.

For ionic structures:

- remember that they are **not** molecules (as there are no covalent bonds)
- the conduction of electricity in ionic structures changes depending on whether the ionic substance is a solid (when the ions are fixed in place) or whether it has been melted or dissolved (when the ions are free to move).

Activity 9: Comparing simple molecules and giant structures

Each statement below describes simple molecules, a giant covalent structure or a giant ionic structure.

Put the letter of each statement in the correct place in the table.

Simple molecules	Giant covalent structures	Giant ionic structures

Statements

A Have strong forces between atoms in a molecule, but weak forces between molecules that only hold them together if it is cold enough

B Have strong forces between all atoms due to the sharing of electrons between the outer shells of atoms

C Have strong forces between ions due to the loss of electrons from one type of atom and gain of electrons by another type of atom

D Some conduct electricity if they have free electrons between layers of the molecules

E Do not conduct electricity as the electrons are tightly bonded between the atoms

F Only conduct electricity when dissolved or melted, as the electrically charged ions can move

Now that you have completed the comparison table, look back at your Frayer maps in Activity 8 to see if there are any further statements that you wish to add about large covalent structures and large ionic structures.

Look back at the concept map that you made at the start of this chapter. Check that you have used all of the key words.

Look at your connections. Are you happy with the connections you made previously? Are there any new connections you can make to show that you have learned something new or deepened your understanding?

TIPS FOR SUCCESS

Go back over the work you have done in this chapter to remind yourself of all the information you have covered. When you are ready, complete this short test.

As you work through it, you can help yourself by:

● reading each question carefully – check you understand the question
● looking for key words to use in your answer
● answering the question in your mind first, before you write it down
● making sure you use correct scientific vocabulary in your answers
● using a piece of spare paper to draft any extended answers first, then when you are happy with it you can write your answer in this book
● checking your answers to make sure that you do not want to make any changes.

Revision test

1   Use the words in the list to add labels to this model of an atom.   [2 marks]

electron   empty space   neutron
orbiting electrons   orbits of the electrons   proton
nucleus containing protons and neutrons

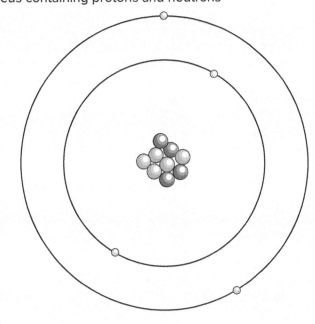

2 Complete the following table by ticking the correct column to show whether each statement is true or false. [3 marks]

Statement	True	False
Atoms with one outer electron are stable.		
Atoms will gain or lose electrons or share them to become stable.		
Atoms always lose electrons to become ions.		

3 Complete the following table by ticking the correct column to show whether each statement is true or false. [3 marks]

Statement	True	False
Each covalent bond is a shared electron between atoms.		
Each covalent bond is a pair of shared electrons between atoms.		
A double bond is where there are two pairs of shared electrons.		

4 Complete the following table by ticking the correct column to show whether each statement is true or false. [3 marks]

Statement	True	False
Atoms that lose an electron will become negatively charged.		
Positive ions are called cations and negative ions are called anions.		
Atoms that gain two electrons will become positive.		

5 Choose the **false** statement. Circle one answer. [1 mark]
   A The nucleus of an atom is positively charged.
   B The electrons are negatively charged.
   C The protons orbit around the outside of the nucleus.
   D An ion is an atom that no longer has a balance of positive and negative charges.

6 Choose the **false** statement. Circle one answer. [1 mark]
   A Covalent bonding involves sharing electrons between atoms.
   B Ionic bonding involves some atoms gaining or losing electrons.
   C Covalent bonds are weak and easy to break.
   D Covalent molecules can be large or small.

7 Draw lines to match the names of the elements below to the colour of circle that would be used to represent that element in the CPK colouring system. [5 marks]

nitrogen	
fluorine/chlorine	
oxygen	
carbon	
hydrogen	

8 Use your knowledge of the structure of molecules to explain why simple molecules have much lower melting and boiling points than giant covalent molecules. [3 marks]

....

....

....

....

....

....

9 Sodium chloride is more commonly known as table salt. It is an ionically bonded substance, made from positive sodium ions and negative chloride ions.
Explain why solid sodium chloride does not conduct electricity. [3 marks]

....

....

....

....

10 Magnesium has two electrons in its outer shell and chlorine has seven electrons in its outer shell. Both atoms would either need a full outer shell (or an empty outer shell) to be stable. They react together to form ions when they make the substance magnesium chloride.
a) For each atom, predict whether it becomes positive or negative as it becomes an ion. Circle the correct answer. [2 marks]
Magnesium: positive / negative   Chlorine: positive / negative
b) How many electrons does each atom lose or gain? Circle the correct answer and complete the gap. [2 marks]

Magnesium loses / gains .... electron(s).

Chlorine loses / gains .... electron(s).

c) Predict the formula of magnesium chloride. [1 mark]

....

Chapter 8 Density

CHAPTER INFORMATION

This chapter will help you to revise your learning about density. By the end of this chapter you should be able to:

- describe what is meant by 'density'
- calculate the densities of solids, liquids and gases
- compare the densities of different materials
- use density to explain floating and sinking
- explain why hot-air balloons float.

REVISION APPROACH: HEXAGON MAP

To help you remember the key words in this topic, you are going to make key word cards. However, this time the cards will be hexagonal. Hexagons tessellate, which means that you can place your hexagon-shaped cards next to each other and make lots of links.

Activity 1: Hexagon key word cards

a) Write each of the following key words on the front of a hexagon-shaped card. Add any images that will help you remember what the word means. If you are unsure about how to make the cards, go back to the beginning of Chapter 5 for a reminder.

centimetres	mass	sink
density	measuring cylinder	top-pan balance
float	metres	volume
grams	metres cubed	
kilograms	round-bottomed flask	

b) On the back of each card, write a definition. Add as many other ideas as you can that link to the key word on the front.
c) Do not worry if you do not know all the key words yet. You can create more cards as you work through this chapter.

Activity 2: Linking hexagon cards

a) Group the hexagons according to the information on each card. Now arrange the hexagons on a sheet of paper or card so that they fit together. Remember, the information on each hexagon must link to the hexagons around it. This is to help you organise what you know. As you work through this chapter, add new cards to your arrangement, even if it means moving some of the cards you have already put down.
b) Take a photograph of the way you have laid out your hexagons. Keep this photograph and return to it later, to see if you want to move the cards to make new or stronger connections between the hexagons.
c) Work with someone else. Choose two or three of their hexagons and ask them to explain in detail why they have placed them next to each other. What are the links between them?

REMEMBER: MASS

Mass is the measurement of the amount of matter in an object. It is measured in kilograms (kg), or for smaller masses in grams (g). In a laboratory or in school, mass is usually measured with a top-pan balance.

1 kg = 1000 g

Mass is different from weight, which is the force of attraction caused by the gravitational pull of the Earth on an object. This force acts towards the centre of the Earth.

Often people say they are 'weighing' an object when they mean they are measuring the mass.

REMEMBER: CALCULATING DENSITY

The density of a substance is a measure of the amount of matter that is present in a certain volume of the substance. Density is calculated using the following equation:

$$density = \frac{mass}{volume}$$

If a substance has a large mass packed into a small space, it has a high density. A substance with a lower mass or where the mass is spread out over a large volume will have a lower density.

You might remember that the basic SI unit of density is found by dividing the unit of mass (kg) by the unit of volume (m^3), so it is kg/m^3. This is pronounced 'kilograms per metre cubed'.

TIPS FOR SUCCESS

Remember to:

- show your working
- lay out your calculation clearly
- put in your units of measurement
- make your answer clear, for example, by underlining it.

MODEL ANSWER

Here is a model answer for calculating the density of an object.

A pallet of bricks has a mass of 1900 kg. The total volume of all of the bricks is $1\,m^3$.

Calculate the density of the bricks.

$$density = \frac{mass}{volume}$$

$$= \frac{1900\ kg}{1\ m^3} = 1900\ kg/m^3$$

We can repeat the calculation for a single brick.

A single brick has a mass of 1.9 kg. The total volume of the brick is $0.001\,m^3$.

$$density = \frac{mass}{volume}$$

$$= \frac{1.9\ kg}{0.001\ m^3} = 1900\ kg/m^3$$

Notice that the density of the bricks is exactly the same, whether you calculate it using the full pallet of bricks or a single brick. This is because density is a property of the material and the same material will have the same density, regardless of the size or shape of the object.

Activity 3: Calculating densities of the same material

Three learners are discussing something that their teacher had asked in class. The teacher asked the class to calculate the density of three different blocks of expanded polystyrene foam, often used as a packaging material.

George said to Lani: 'I think the large block will be more dense than the small block.'

Lani disagreed, saying: 'I think the small block will be more dense than the large block.'

Mass is usually measured in grams (g) or kilograms (kg).

1 kg = 1000 g

To convert from kg to g, multiply by 1000.

To convert from g to kg, divide by 1000.

Length is usually measured in millimetres (mm), centimetres (cm) or metres (m).

1 m = 100 cm
    = 1000 mm.

To convert from mm to cm, divide by 10.

To convert from cm to m, divide by 100.

Remember, it is much easier to convert the individual length measurements than convert the volume.

Theo said: 'I think the density of all three blocks will be the same, as it is the same material.'

Use the information in the table to calculate the densities of the blocks and work out who is correct.

Block	small	medium	large
Mass	0.02 kg	1 kg	20 kg
Volume	0.001 m³	0.05 m³	1 m³
Density			

Activity 4: Calculating densities of different materials

Use the information in the table to calculate the density of each of the following substances. Remember to show your working in the space provided.

Substance	ice	steel	air
Mass	917 kg	8 kg	127 kg
Volume	1 m³	0.001 m³	100 m³
Density			

Activity 5: Converting mass and volume

1   Practise converting the following masses into grams.

   a) 50 kg = ...   c) 419 kg = ...

   b) 370 kg = ...

2   Practise converting the following masses into kilograms.

   a) 50 000 g = ...   c) 100 g = ...

   b) 2500 g = ...

3   Practise converting the following lengths. Make sure to notice the units you must convert each measurement into.

   a) 5 cm = ... m   c) 1000 cm = ... m

   b) 350 mm = ... m   d) 20 m = ... cm

REMEMBER: CALCULATING THE DENSITY OF RECTANGULAR SOLID OBJECTS

These are the steps to calculate the density of a rectangular solid block:

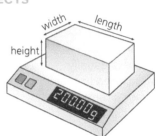

- Find the mass of the block by placing it on a balance. The mass might be recorded in grams or kilograms, depending on the scale.
- Find the volume of the block by multiplying the length, width and height together.
- Divide the mass of the block by its volume.

Activity 6: Density problems

Some learners have been measuring the density of regular blocks of solid materials. All their original measurements are correct, but they have made some mistakes in their calculations and conversions.

Identify their mistakes and give them some advice as a correction.

Material	Mass of block	Dimensions	Density	Correction
wax	7.6 g	2 cm	0.95 kg/m³	
		2 cm		
		2 cm		
aluminium	0.162 kg	5 cm	0.00027 kg/m³	
		4 cm		
		3 cm		
stone	81 g	30 mm	0.003 g/cm³	
		30 mm		
		30 mm		

Activity 7: Measuring density

Some learners have been carrying out an experiment in class. They have written up their experiment in their notebooks but have made some mistakes in their writing.

Read through their description of the experiment and see if you can spot the mistakes. Once you have found a mistake, write a correction in. There is one mistake on every line; they have been separated out to help you.

We started by getting the weight of the block using a top-pan balance.

....

We found the volume of the block by measuring the depth and length using a ruler.

....

We calculated the volume of the block by multiplying the depth and length together.

....

REMEMBER: CALCULATING THE DENSITY OF AN IRREGULARLY SHAPED OBJECT

When calculating the density of an irregularly shaped solid, like a pebble, the calculation for density remains the same: mass divided by volume. To measure the mass, we can place the pebble on a balance. However, the method for measuring the volume is different. We cannot easily calculate an object's volume when it is an irregular shape.

Instead, we use the idea of displacement – the volume of liquid that is displaced when the solid object sinks in water. The method for this is:

- Fill a measuring cylinder half-full of water. Take a reading of how much water it contains.
- Lower the object (in this case, a pebble) into the water.
- Take a second reading to measure the water level again.
- To work out the volume, subtract the second reading from the first.

measuring cylinder

second reading

first reading

water

pebble

Activity 8: Calculating the volume of irregularly shaped objects

Use the information in the table to calculate the missing values.

Volume of liquid at start	Volume of liquid + object	Volume of object
50 cm³	60 cm³	
20 cm³		27 cm³
	85 cm³	12.5 cm³

REMEMBER: DENSITY OF LIQUIDS

Some liquids are denser than others. If the liquids do not mix, the less dense liquid will float in a layer above the more dense liquid. You can use this idea to compare the density of the layers.

Activity 9: Oil and water

Jamal is making a vinaigrette salad dressing by combining oil and vinegar. During the making of the vinaigrette, Jamal notices that the oil forms a layer on top of the vinegar.

Which liquid is more dense: oil or vinegar? Explain your answer.

....

....

REMEMBER: CALCULATING THE DENSITY OF LIQUIDS

To calculate the density of a liquid, we need the same information as when calculating the density of a solid: mass and volume. We can easily measure the volume of liquids with a measuring cylinder. The challenge is measuring the mass. We can't just pour the liquid on to a balance as it would run away.

Instead, we need to use a container to hold the liquid, so we can find the mass using the balance. We also need to know the mass of the container so that we can subtract that from the total mass of the liquid plus the container.

MODEL ANSWER

A container of washing-up liquid has a mass of 183 g. The mass of the empty container is 80 g.

The volume of washing-up liquid is 100 cm³. Calculate the density of the washing-up liquid in g/cm³.

Step 1: Find the mass of the washing-up liquid.

mass of washing-up liquid = mass of liquid and container − mass of container
$$= 183 \text{ g} − 80 \text{ g} = 103 \text{ g}$$

Step 2: Calculate the density of the washing-up liquid.

$$\text{density} = \frac{\text{mass}}{\text{volume}}$$

$$= \frac{103 \text{ g}}{100 \text{ cm}^3} = 1.03 \text{ g/cm}^3$$

REMEMBER: CALCULATING THE DENSITY OF GASES

It is difficult to find the mass of a gas using a balance as it would escape. We have to use a similar method to the one for a liquid to find the mass of the gas, which works for finding the mass of gases that are more dense than air. When we do this, we also need to check the volume of the container.

This is how to find the density of a gas:

- Measure the mass of the container and gas on a sensitive balance.
- Remove the gas from the container.
- Measure the mass of the empty container.
- Calculate the mass of the gas by subtracting the mass of the container from the total mass of the container and gas.
- Find the volume of the container by filling it with water and measuring the volume of the water using a measuring cylinder.

Activity 10: Checking a method

A group of learners have been measuring the density of a gas in their school laboratory. They have used a very sensitive balance to find the mass of the gas in its container.

They have calculated the density of the gas using the measurements obtained in the experiment.

They compare the density they calculated with what the density of the gas should be. They realise that the density figure they have calculated is too high.

Look through their method and calculation and see if you can spot their mistake(s).

Method:

- First, we kept the temperature of the gas close to 0 °C and the pressure of the gas was similar to atmospheric pressure.
- The volume of the flask was marked as 1000 ml. We converted this to give a volume of 0.0013 m³.
- We zeroed the balance and placed the flask on the balance. The mass was 394.02 g, which equals 0.39402 kg.
- To calculate the density, we divided the mass in kg by the volume in m³. This gave us a density of 394.02 kg/m³, which was much higher than we expected. The density should be 2.90 kg/m³.

What mistake(s) did they make? How can they put it right?

....

....

....

> **REMEMBER: GAS DENSITY DEPENDS ON TEMPERATURE AND PRESSURE**
>
> Remember that the effect of temperature on the density of gases will be much more noticeable than for most liquids and solids. This is due to the larger particle movements involved in gases.
>
> We use a standard temperature (0 °C) and pressure (similar to atmospheric pressure) to compare the densities of gases.
>
> Increasing the temperature will increase particle motion and tend to reduce the density of the gas. Increasing the pressure will force particles more closely together and tend to increase the density of the gas.

Activity 11: A hot-air balloon

Imagine that you have been asked by your teacher to investigate making hot-air balloons using hairdryers and plastic bags. Your teacher has asked you to explain how they work.

The explanation has been jumbled up. The first and last statements have already been listed in the sequence to help you.

A The hairdryer transfers electrical energy into thermal (heat) energy. This warms the surrounding air.

B The less dense warm air floats above the more dense cool air.

C As the air warms, the particles move around more. This means that they spread out more.

D Warm air is blown from the hairdryer into the plastic bag, warming the air in the bag.

E As the particles are now more spread out, the air is less dense than normal.

F The plastic bag rises upwards.

The correct sequence is: A, ..., ..., ..., ..., F.

Activity 12: Revisit hexagons

Go back to your hexagon word cards from Activity 1. Having completed this chapter, lay out the hexagons again, making connections between the cards. Compare this to the way you laid them out the first time. Are you happy with the original connections? What new connections have you made?

Revision test

1 Which equation is used to calculate density? Circle the correct answer. [1 mark]

  A Density = mass × volume   B Density = $\dfrac{\text{volume}}{\text{mass}}$

  C Density = $\dfrac{\text{mass}}{\text{volume}}$

2 Which equation is used to calculate the volume of a regular rectangular solid? Circle the correct answer. [1 mark]

  A Volume = width × height   B Volume = width × height × length

  C Volume = $\dfrac{\text{width × height}}{\text{length}}$

3 What do you do to convert from g to kg? Circle the correct answer. [1 mark]

  A Divide by 1000   B Multiply by 1000

  C Multiply by 1000000

4 What do you do to convert from m^3 to cm^3? Circle the correct answer. [1 mark]

  A Divide by 100   B Multiply by 100

  C Multiply by 1000000

5 Use the data listed below to calculate the density of a regular solid block of cheese. [2 marks]

  width = 8 cm   length = 9 cm

  height = 22 mm   mass = 170 g

6 Calculate the density of this stone using the following information from an experiment. [2 marks]

  mass of stone = 80 g   volume of liquid = 40 cm^3 volume of liquid + stone = 90 cm^3

7 A container full of honey has a mass of 235 g. The mass of the empty container is 33 g. The volume of the honey is 144 cm^3. Calculate the density of the honey in g/cm^3. [2 marks]

8 When measuring the volume and mass of a gas, state three things you need to remember.   [3 marks]

....

....

....

9 A teacher sets up three liquids in a container: oil, water and sugar syrup. The liquids do not mix but instead settle into three layers, as shown in the image below. The teacher then drops three objects into the container:
   – a steel bolt   – a tomato   – a sultana.

The objects settle into the container as shown.

The density of oil = 920 kg/m^3
The density of water = 1000 kg/m^3
The density of sugar syrup = 1430 kg/m^3

Estimate a value for the density of each object.

Steel bolt density estimate ....

Tomato density estimate ....

Sultana density estimate ....

Explain your reasoning behind your estimates.   [4 marks]

....

....

....

....

....

....

....

....

Chapter 9 Displacement reactions

REMEMBER: THE REACTIVITY SERIES OF METALS

Scientists sort metals into a reactivity series. The table shows some of the metals in the series, starting with the most reactive and ending with the least reactive.

The reactivity series provides a clue to how metals will be displaced when chemical reactions take place between them.

Although hydrogen is not a metal, it is sometimes included as a reference. This helps us to decide if the metals will react with acids or water (hydrogen is a common element to both acid and water).

If the metal is above hydrogen in the series, the metal will displace the hydrogen in an acid. The most reactive metals will also react quickly (and sometimes explosively) with water (which also contains hydrogen), such as the Group 1 metals. You might remember seeing very fast reactions between metals like potassium and water.

In our table, hydrogen is above copper and below lead. This means that it is more reactive than copper, but less reactive than lead.

more reactive ↑

Metal	Symbol
potassium	K
sodium	Na
calcium	Ca
magnesium	Mg
aluminium	Al
zinc	Zn
iron	Fe
lead	Pb
(hydrogen)	(H)
copper	Cu
silver	Ag
gold	Au

increasing in reactivity

less reactive

Activity 1: Using the reactivity series

For each of the following pairs of metals, circle the one that is more reactive. Use the reactivity series to help you decide.

a) Calcium or silver
b) Iron or copper
c) Magnesium or gold
d) Magnesium or iron
e) Aluminium or zinc

potassium	people
sodium	squeezing
calcium	cheese
magnesium	means
aluminium	a
zinc	zombie
iron	in
lead	light
(hydrogen)	houses
copper	can
silver	smell
gold	good

Activity 2: Remembering the reactivity series

It is helpful to know the reactivity of some common metals so that you do not have to refer back to the reactivity series every time.

One way is to use a mnemonic. This means you take the first letter of the names of the metals in the reactivity series and then come up with a memorable phrase that uses those same letters.

The table shows an example. Try to make a better one that helps you remember the order of the letters. Once you can do that, check that you can then match the names of the metals in the series with the letters.

Activity 3: Key word cards

Later in this topic, you will complete a Frayer map (look back at Chapter 7 to see an example of how these should look) with the phrase 'displacement reactions' in the middle. To do this, you will need to be familiar with some key words and phrases.

To help familiarise yourself with the key words, make a series of key word cards for the following words and phrases:

acid	metal salt	reactivity series of
displacement	product	metals
hydrogen	reactant	thermit reaction
metal		

On the front of each card, write the word or phrase. On the back, write a definition and any key facts about the phrase.

REMEMBER: DISPLACEMENT REACTIONS

In a displacement reaction, a less reactive element in a compound is replaced by a more reactive one. Scientists use the word 'displaced' instead of 'replaced', and this word gives the name to the reaction.

As we will see later in this chapter, not all combinations of chemicals will result in a displacement reaction. Whether it will take place depends on whether the metal involved is more reactive than the element it is reacting with. The elements involved in displacement reactions are often metals, which make up most of the elements in the periodic table. However, sometimes it is another element, like hydrogen.

MODEL ANSWER

Adding zinc powder to copper sulfate solution will result in a displacement reaction, as zinc is more reactive than copper.

We can show this as a chemical word equation:

zinc + copper sulfate → zinc sulfate + copper

If we added copper powder to zinc sulfate solution, no displacement reaction would take place. This is because copper is less reactive than zinc.

copper + zinc sulfate → **no reaction**

Remember, the signs that a chemical reaction has taken place include:

- changes of colour
- fizzing or bubbling
- changes of temperature, for example:
  - becoming hotter (a sign of an exothermic reaction that releases stored energy as thermal energy into the surroundings)
  - becoming colder (a sign of an endothermic reaction that takes in thermal energy from the surroundings).

Activity 4: Displaced or not?

Some learners have been carrying out experiments to check whether a displacement reaction takes place. Their results are shown in the table.

Use the observations to decide whether a displacement reaction has taken place.

Reactants	Observations	Has a reaction taken place? (Circle the correct answer.)	Give a reason for your answer
magnesium + copper chloride	bubbles and feels warm to the touch, solution seems paler and red/brown solid lumps appear	yes / no	
iron + zinc nitrate	no change of colour, no bubbles or change of temperature	yes / no	
zinc + lead nitrate	silvery metal becomes dull and grey	yes / no	

REMEMBER: THE DISPLACEMENT OF HYDROGEN BY MAGNESIUM

Magnesium is above hydrogen in the reactivity series, so displaces it. When magnesium is added to water, the magnesium slowly displaces the hydrogen in the water molecules and forms magnesium hydroxide, releasing hydrogen gas. As the reaction is quite slow, patience is needed to observe the bubbles of hydrogen forming.

As magnesium hydroxide is colourless, we use indicator to detect it. Magnesium hydroxide is a weak alkali. It turns the indicator phenolphthalein from colourless to pink. The colour change in the right-hand tube in the picture shows that magnesium hydroxide has formed.

Activity 5: The displacement of hydrogen by magnesium

a) How can you tell that a reaction has taken place when magnesium is added to water?

....

....

....

b) Use the information about the reaction to complete this word equation for the displacement reaction that takes place:

magnesium + water → .... + ...

REMEMBER: THE REACTION OF METALS WITH ACIDS

When metals react with acids, they displace hydrogen from the acid and form a salt solution. You will find out more about salts and how they are made in the next chapter. The general word equation for this reaction is:

metal + acid → metal salt + hydrogen

For example: zinc + hydrochloric acid → zinc chloride + hydrogen

In this reaction, you would make the salt (zinc chloride) and a gas (hydrogen). During the reaction you would see bubbles of hydrogen gas rising through the solution.

Zinc chloride dissolves in water easily and is colourless when dissolved. In order to see the crystals of the salt, you would have to filter out any zinc that had not reacted, using a funnel and filter paper. You would then have to evaporate the water from the solution (carefully, as it may still be acidic and heating an acid can create dangerous acidic fumes). Once the water had evaporated, you would see small white crystals of zinc chloride.

Remember that not all metals will react with acids. A metal will only react with acid if it appears above where hydrogen would be in the reactivity series.

Activity 6: Making chemical equation cards

The chemical equations given here are examples of very similar reactions that you will see over and over again. It is really helpful to learn these reactions so that you can recall them quickly. You also need to be able to spot the different parts of the reaction quickly. To help you to do this, make some reaction cards.

Start by making cards for the general equation:

metal + acid → metal salt + hydrogen

- You will need individual cards for the metal, acid, metal salt and hydrogen. You can also make cards for the '+' signs and the arrow.
- Use a colour code. For example, all metals could be in grey or silver, acids could be in red, hydrogen could be in orange and the salt could be partly in grey/silver for the metal and partly in another colour for the salt.
- You can also add pictures to help you remember. Perhaps you could come back to these cards in the later chapters and add diagrams of some of the atoms and molecules involved.

- Once you have these cards, test yourself by laying them out in the right order from memory. Come back a few days later and try again.
- You can also make new sets of cards for each type of metal, acid and salt. So, in our example above you can make extra cards for zinc (as an example of a metal), hydrochloric acid (as an example of an acid) and zinc chloride (as an example of a salt).

Activity 7: Reactions with acid?

Using the reactivity series, predict whether the following metals will react with hydrochloric acid or not. Give a reason for each answer.

a) Iron ....

....

b) Copper ....

....

c) Magnesium ....

....

d) Zinc ....

....

Activity 8: Writing an equation for metals and acids

Write a chemical word equation to show the reaction for a metal from Activity 7 that you think will react with hydrochloric acid.

Clue: When hydrochloric acid reacts with a metal, the salt formed is a metal chloride.

.... + ... → .... + ...

REMEMBER: THE THERMIT REACTION

One particularly useful displacement reaction is the thermit reaction. This is often used to join railway tracks together securely. The rails are heated, then iron oxide powder and aluminium powder are added. The heat from the rails causes a displacement reaction in which the aluminium displaces the iron, producing even more heat. This melts the iron at the ends of the rails and joins them together in a process called welding.

The prefix 'therm' relates to temperature – you can remember that a thermit reaction is a hot one by looking out for the 'therm' prefix. 'Therm'-it reactions release a lot of 'therm'-al energy and create a high temperature that you would measure with a 'therm'-ometer.

Activity 9: Joining rails

Railway workers are trying to join some rails on a railway line using the thermit reaction.

a) Complete the equation to show the displacement in the thermit reaction:

aluminium + iron oxide → .... + ...

Due to a recent spell of hot weather, the number of rail repairs was higher than normal. This meant that the team has run low on supplies of aluminium powder and they do not have enough to complete all of the repairs.

One of the workers suggests that they replace the aluminium powder with another metal so that they can make the repairs. She suggests using lead power, which looks similar to aluminium.

b) Would replacing aluminium with lead allow the thermit reaction to take place? Using the reactivity series, explain why or why not.

....

....

....

....

REMEMBER: DISPLACEMENT OF METALS

When metals react with acids, they displace hydrogen from the acid and form a salt solution. In a similar way, a more reactive metal can displace a less reactive metal from a salt solution of the less reactive metal.

For example, adding zinc powder to copper sulfate solution will result in a displacement reaction, as zinc is more reactive than copper. We can show this with a word equation:

zinc + copper sulfate → zinc sulfate + copper

Activity 10: Copper plating experiment

Some learners were trying to coat objects made out of different metals with a decorative coating of copper as part of an art project.

To coat the object, they carefully scrubbed the surface clean with wire wool. They then placed each object in a solution containing copper sulfate (a metal salt). They left the object for a while to see if the metal in the object displaced the copper in the copper sulfate solution to leave a coating of copper.

They found that some of the objects became coated but others did not.

a) For each object, predict whether a reaction to make a copper coating would occur. If so, write down the word equation for the displacement reaction. If not, write 'no reaction'.

1 An iron key ....

2 A silver spoon ....

3 A gold ring ....

4 A zinc name badge ....

b) One of the learners suggests that they have not left the objects in the solution long enough and that if they leave them longer, then all the objects will be coated. Is this suggestion correct? Explain your answer.

....

....

....

c) Write down a rule that would help the learners decide what type of objects to use for their project.

....

....

....

Activity 11: Make a Frayer map

At the start of this chapter, you made a set of key word cards (Activity 3). Use your learning from the key word cards and other activities to help you complete a Frayer map for displacement reactions. Copy out the Frayer map template from Chapter 7 on a separate piece of paper. Check that you are happy with what you have written on each of your key word cards, making any corrections that you need.

Now complete your Frayer map, ensuring that you have:

- a clear definition
- clear and accurate characteristics, features and facts
- several examples
- any non-examples.

TIPS FOR SUCCESS

Go back over the work you have done in this chapter to remind yourself of all the information you have covered. When you are ready, complete this short test.

As you work through it, you can help yourself by:

- reading each question carefully – check you understand the question
- looking for key words to use in your answer
- answering the question in your mind first, before you write it down
- making sure you use correct scientific vocabulary in your answers
- using a piece of spare paper to draft any extended answers first, then when you are happy with it you can write your answer in this book
- checking your answers to make sure that you do not want to make any changes.

Revision test

1 Complete the following table by ticking the correct column to show whether each statement is a sign of a displacement reaction taking place. [4 marks]

Statement	Yes	No
A piece of silver metal in a blue solution remains as a piece of silver metal in a blue solution.		
A silver/grey powder turns into a brown powder.		
A green solution loses its colour and becomes colourless when a metal is added.		
The test tube containing the reactants heats up on its own.		

2   Complete the following table by ticking the correct column to show whether each
    statement is true or false.   [4 marks]

Statement	True	False
If one metal is more reactive than another, it will displace the less reactive metal in a compound.		
Metals towards the bottom of the reactivity series are always silver in colour.		
If a metal is less reactive than hydrogen, it will always react with acids.		
Magnesium will displace any other metal.		

3   Which of the following metals will displace the iron from iron chloride? Use the reactivity
    series to help you answer this question. Circle the correct answer(s).   [2 marks]
    A Gold   B Calcium   C Lead   D Zinc

4   Which of the following metal salts will react with zinc in a displacement reaction? Use the
    reactivity series to help you answer this question. Circle the correct answer(s). [2 marks]
    A Calcium chloride   C Potassium sulfate
    B Iron nitrate   D Copper sulfate

5   Complete the word equations for the following displacement reactions, if you think a
    reaction will take place. If you think there will be no reaction, write 'no reaction'. Use the
    reactivity series to help you.   [4 marks]

    a) calcium + zinc sulfate → ....

    b) lead + iron chloride → ....

    c) zinc + hydrochloric acid → ....

    d) copper + nitric acid → ....

6   A group of learners notice that the metal tin is not included in the reactivity series that they
    have been given. Describe an experiment using a displacement reaction that they could
    use to work out where tin should be in the reactivity series. In your answer, include:
    – the method (what they would do)
    – the observations they would make
    – what would tell them if there was a reaction
    – how they would decide if it was more or less reactive.   [4 marks]

    ....

    ....

    ....

    ....

    ....

    ....

    ....

    ....

    ....

    ....

Chapter 10 Preparing common salts

CHAPTER INFORMATION

This chapter will help you to revise your learning about making common salts. By the end of this chapter you should be able to:

- describe how salts are made with acids
- use word equations to describe reactions
- describe how a substance can be purified through evaporation, crystallisation and filtration
- describe how to prepare a salt from a metal and an acid
- describe how to prepare a salt from a metal carbonate and an acid

Activity 1: Key word cards

Later in this topic, you will complete a Venn diagram about the two ways that you can prepare common salts (acid + metal and acid + metal carbonate). This will help you understand the similarities and differences between the two techniques.

To help familiarise yourself with the key words in this topic, make a series of key word cards for the following words and phrases:

acid	filtrate	metal carbonate	residue
carbon dioxide	filtration	metal salt	soluble
crystallisation	hydrogen	mixture	solution
evaporation	insoluble	product	water
filter	metal	reactant	

On the front of each card, write the word or phrase. On the back, write a definition and any key facts about the phrase.

Some of these cards are similar to those in Chapter 9; you might choose to reuse those if you have made them already, adding new information.

REMEMBER: SALTS

There is more to salts than the salt on our table. The proper chemical name for the salt we use in our cooking is sodium chloride. It is just one example of a salt and there are many others.

Salts have large ionic structures, a phrase you might recognise from Chapter 7. This means that the structures are made of ions and so they have very high melting points. It also means that many salts can dissolve in water and separate out into their ions.

Some salts, like sodium chloride, are white in colour and colourless when dissolved. Other salts are coloured (often brightly), as you can see from the picture. They also form brightly coloured solutions if they dissolve.

Metal salts have two parts to their names: the metal and the type of salt. There are three common types of salt:

- chlorides
- sulfates
- nitrates.

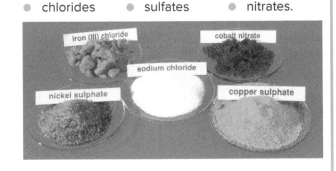

Activity 2: Sorting out substances

The following chemical substances have been jumbled up.

zinc   nitric   iron sulfate   magnesium chloride
hydrochloric   calcium   sulfuric   lead nitrate   copper

Write them into the table below, identifying three metals, three acids and three salts.

Metals	Acids	Salts

REMEMBER: ACIDS AND THEIR SALTS

You might remember from Chapter 9 that you can make a salt by reacting a metal with an acid, as long as the metal is reactive enough. This happens when the metal displaces the hydrogen from the acid.

The general equation for making a salt in this way is:

metal + acid → metal salt + hydrogen

For example: zinc + nitric acid → zinc nitrate + hydrogen

Activity 3: Reactions with acid to make metal salts

Complete the sentences to show what salts are produced with each type of acid.

a) Hydrochloric acid reacts with metals to produce .... salts.

b) Sulfuric acid reacts with metals to produce .... salts.

c) Nitric acid reacts with metals to produce .... salts.

REMEMBER: DIFFERENT ACIDS, DIFFERENT SALTS

Different types of acid produce different types of salt when reacted with metals. The three main acids that are used to make salts are hydrochloric acid, sulfuric acid and nitric acid. The salts produced by these acids are chlorides, sulfates and nitrates.

This is a type of displacement reaction. All acids contain hydrogen. They also contain a second part, which forms the metal salt. A metal that is more reactive than hydrogen in the reactivity series will displace the hydrogen, which is released as a gas. The metal will bond with the other part of the acid to form a metal salt.

Use the list of substances below to match up the metals, acids and salts across each row of the table.

magnesium   sulfuric acid   hydrochloric acid
magnesium nitrate   zinc sulfate   calcium

Metal	Acid	Salt
	nitric acid	
		calcium chloride
zinc		

REMEMBER: METAL CARBONATES

Another way of making salts is to add an acid to a chemical called a carbonate. A carbonate is a chemical that contains two elements: carbon and oxygen. A good way to remember this is to think of fizzy drinks – you may have heard fizzy drinks being referred to as 'carbonated'. This is because they have had carbon dioxide gas added to them, which is made up of carbon and oxygen.

It is also a reminder that carbonates release carbon dioxide when you make a salt:

metal carbonate + acid → salt + carbon dioxide + water

For example:

copper carbonate + sulfuric acid → copper sulfate + carbon dioxide + water

As the chemicals involved are solutions, the carbon dioxide released forms bubbles.

It is really helpful to learn common reactions so that you can recall them quickly. You also need to be able to spot the different parts of the reaction quickly. To help you to do this, make some reaction cards.

Start by making cards for the general equation:

metal carbonate + acid → salt + carbon dioxide + water

- You will need individual cards for the metal carbonate, acid, metal salt, carbon dioxide and water. You can also make cards for the '+' signs and the arrow.
- Use a colour code. For example, the metal part of the metal carbonates could be in grey or silver and the carbonate part could be in black. Acids could be in red, carbon dioxide could be in black, water in blue and the salt could be partly in grey/silver for the metal and partly in another colour for the salt.
- You can also add pictures to help you remember. Perhaps come back to these cards after the later chapters and add diagrams of some of the atoms and molecules involved.

- Once you have these cards, test yourself by laying them out in the right order from memory. Come back a few days later and try again.
- You can also make new sets of cards for each type of metal carbonate, acid and salt. For example, you could make extra cards for zinc carbonate (as an example of a metal carbonate), hydrochloric acid (as an example of an acid) and zinc chloride (as an example of a salt).

REMEMBER: EVAPORATION AND CRYSTALLISATION AS PURIFICATION TECHNIQUES

One way of making crystals in the preparation of salts is to start with a concentrated solution of a substance. When the concentrated solution is gently heated in an evaporating dish, the solvent begins to evaporate. When it is almost dry, the heat is removed and the remainder of the solvent evaporates slowly, leaving behind crystals. The pure salt crystals can then be dried by patting them gently with a paper towel. When making salts in the lab, the solvent is usually water.

Activity 6: Describing evaporation and crystallisation

A teacher describes the process of evaporation as a type of purification. Three learners are discussing this in class.

Ahmed says: 'This isn't purification. You have exactly the same substance as you did at the start, so there is no chemical reaction taking place.'

Aga says: 'This is a type of purification, as you are separating the salt from the water.'

Emilia says: 'This is definitely a chemical reaction, as you can see the bubbles when the water evaporates.'

Which of the statements do you most agree with? Justify your answer.

....

....

....

....

....

....

REMEMBER: FILTRATION AS A PURIFICATION TECHNIQUE

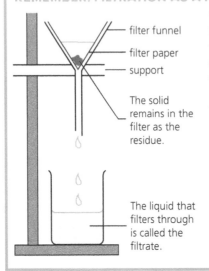

filter funnel

filter paper

support

The solid remains in the filter as the residue.

The liquid that filters through is called the filtrate.

In many laboratory experiments, filtration is used to separate out pure substances from a mixture. It works when one of the substances is insoluble in a solvent and the other is soluble, such as excess metal in a reaction between a metal and acid. The diagram shows how filtration works.

Activity 7: Filtration

Using the information about filtration above, write the letter of each statement in the correct order to describe the process of filtration.

A The solid particles are left behind on the paper.
B A piece of filter paper is folded to make a cone.
C The folded paper is inserted into a filter funnel.
D During filtration, only the liquid passes through the holes in the filter paper because they are so small.
E The funnel is supported above a collecting vessel and the mixture to be separated is poured into the funnel.

REMEMBER: PREPARING A SALT FROM A METAL AND AN ACID

When carrying out the reaction to prepare a salt from a metal and an acid, scientists often use an excess of metal (in other words, too much metal compared to the acid). This ensures that all the acid reacts to leave a pure sample of the salt at the end.

The excess metal needs to be separated out from the solution by filtration (as seen in the previous section), after the reaction is complete but before the sample of salt is prepared.

Once the salt has been produced, it needs to be separated from the solution by evaporation and crystallisation.

Activity 8: Experiment to make salt using acid + metal

Some learners have been carrying out an experiment to make a sample of zinc sulfate salt using the method of acid + metal.

a) Write out the word equation for this reaction.

....

Their experimental method and some observations are below. Unfortunately, they have made two mistakes in their method and the method and observations are jumbled up, so are not in the correct order.

A   The zinc metal remains behind and the liquid passes through into a beaker.

B   Granulated zinc is added to nitric acid in a flask.

C   Eventually, the bubbles are no longer produced and some magnesium metal remains in the flask.

D   The contents of the flask are then poured onto filter paper in a filter funnel.

E   Bubbles of gas rise from the metal, pass through the liquid and escape into the air.

b) Write down the correct sequence of letters.

....

c) What are the two mistakes? In each case, describe the correction.

Mistake 1 ....

Correction 1 ....

Mistake 2 ....

Correction 2 ....

> **REMEMBER: PREPARING A SALT FROM A METAL CARBONATE AND AN ACID**
>
> A similar procedure is used to make salts from metal carbonates and acids. In this case, rather than an excess of metal, the reaction uses an excess of metal carbonate. It still uses the processes of filtration and evaporation in the same order. It also produces a gas, although the gas in this case is carbon dioxide instead of hydrogen.

Activity 9: Experiment with some substitution to make a different substance

A teacher asks their class some questions about making magnesium chloride from magnesium carbonate. Magnesium chloride is a soluble salt.

Daanesh's responses to the questions are in the table below – some are correct and some are incorrect. Use your knowledge of making salts to decide which of his answers are correct and which are incorrect. For any that are incorrect, give the correct answer.

Start by writing out the chemical equation for the reaction. Use Activity 3 at the start of this chapter to help you choose the correct acid.

....

Question	Daanesh's response	Is he correct? (Circle the correct answer.)	Correct answer (if needed)
What is the name of the acid used?	nitric acid	yes / no	
What would be left in the filter funnel?	excess unreacted magnesium carbonate	yes / no	
How would you separate the magnesium chloride from the solution?	filtration	yes / no	
A gas is produced. What do you think it is and how would you test for it?	hydrogen gas, tested for by collecting, inserting a lit splint and getting a 'squeaky pop'	yes / no	

REMEMBER: PURIFICATION TECHNIQUES

Different purification techniques are used for different purposes.

As you saw earlier in this chapter, evaporation and crystallisation are used to separate soluble salts from the water in a solution. Filtration is used to separate out insoluble solids, such as metals, from liquids.

Activity 10: Purification techniques

Draw lines to match each purification technique to the situation described in the experiment.

Separating copper powder from zinc sulfate solution	Filtration, evaporation and crystallisation
Obtaining solid iron chloride from an incomplete reaction of copper chloride and iron (filings)	Filtration
Separating soluble magnesium chloride from a solution	Evaporation and crystallisation

REVISION APPROACH: VENN DIAGRAM

Venn diagrams are useful for showing the overlap and differences between ideas. As there are both common themes and differences between the two methods of preparing common salts in this chapter, this is a useful method to compare them.

Activity 11: Making a Venn diagram

a) On a separate piece of paper, draw two overlapping circles. Label the first circle 'Making salts with acid + metal' and the second 'Making salts with acid + carbonate'.
b) Information and diagrams that **only** apply to the first method are placed in the first circle, but **not** within the overlapping area.
c) Information and diagrams that **only** apply to the second method are placed in the second circle, but **not** within the overlapping area.
d) Any information that applies to both methods goes in the overlapping area.

If you are not confident with this topic, you might want to use a pencil so that it is easier to correct later.

TIPS FOR SUCCESS

Go back over the work that you have done in this chapter, to remind yourself of all the information you have covered. When you are ready complete this short test.

As you work through it, you can help yourself by:

● reading each question carefully – check you understand the question
● looking for key words to use in your answer
● answering the question in your mind first, before you write it down
● making sure you use correct scientific vocabulary in your answers
● using a piece of spare paper to draft any extended answers first, then when you are happy with it you can write your answer in this book
● checking your answers to make sure that you do not want to make any changes.

Revision test

1 Complete the following table by ticking the correct column to show whether each statement is true or false.   [4 marks]

Statement	True	False
Zinc is a metal.		
Zinc sulfate is a metal.		
Zinc sulfate is a salt.		
Sulfuric is a type of acid.		

2 Complete the following table by ticking the correct column to show whether each statement is true or false.   [4 marks]

Statement	True	False
Hydrochloric acid is used to make chloride salts.		
Sulfuric acid is used to make sulfate salts.		
Nitric acid is used to make chloride salts.		
Sulfuric acid is used to make nitrate salts.		

3 Choose the **false** statement. Circle one answer.   [1 mark]
   A Evaporation and crystallisation are used to separate salts from their solutions.
   B Evaporation and crystallisation are used to remove excess reactants.
   C Evaporation and crystallisation are a form of purification.
   D Evaporation involves heating.

4 Choose the **false** statement. Circle one answer.   [1 mark]
   A Filtration is used to separate salts from their solution.
   B Filtration is used to remove excess reactants.
   C Filtration is a form of purification.
   D Filtration often uses a folded filter paper, which acts as a mesh to catch solids, allowing liquids to pass through.

5 Complete the following equations: [4 marks]

a) magnesium + nitric acid → ....

b) lead + sulfuric acid → ....

c) calcium carbonate + nitric acid → ....

d) magnesium carbonate + hydrochloric acid → ....

6 A group of learners was asked to make copper nitrate (a salt). To make this, the group has been provided with a selection of acids and samples of the following compounds:

– magnesium carbonate
– calcium carbonate
– copper carbonate.

a) Choose one compound from this list and one acid that the group would need to make copper nitrate. [2 marks]

....

b) Write down the method that they would use, highlighting each step that they would take. [5 marks]

....

....

....

....

....

....

....

....

....

c) What would the word equation for this reaction be? [1 mark]

....

d) What **two** changes would be needed to make calcium chloride instead? [2 marks]

1 ....

2 ....

Chapter 11 Rates of reaction

REVISION APPROACH: POSTER

A poster is a good way to combine ideas that include both words and diagrams. It is a really helpful strategy for this topic, as understanding rates of reaction requires the use of important vocabulary **and** explanatory diagrams to help you remember them more easily.

Activity 1: Revision poster

Before you start the activities, begin by drawing a poster showing what you already know about rates of reaction. On the poster include diagrams, key phrases and key vocabulary. You can complete your poster either by hand on a (large) piece of paper or on a computer.

To help you, a list of key vocabulary is included below. You may also want to look back at other topics on reactions, as some of the key ideas are included in those topics too.

catalyst	energy	product	temperature
closed system	mass	rate	triangle of fire
collision	non-closed	reactant	volume
concentration	system	reaction	word equation
conservation	particle	symbol equation	

REMEMBER: RATES OF REACTION

Chemical reactions take place at different rates. Some reactions have a high rate and take place quickly; for example, in an explosion. Some reactions have a low rate and take place over a much longer time, such as the rusting of iron. We can think of the rate of reaction as the speed at which the reactants make the products in a reaction.

Some things can be used to increase the rate of reaction (speed it up), or decrease the rate of reaction (slow it down). We will explore these ideas in this chapter.

Activity 2: Rank the rates of reaction

Look at the reactions below. Write their letters in order from fastest to slowest.

A   A cake being baked
B   A firework burning then exploding
C   An iron railing rusting
D   An apple's surface turning brown after being bitten into

Fastest .... **Slowest**

REMEMBER: THE CONSERVATION OF MASS

The conservation of mass is a key idea in chemical reactions. Understanding this can help you make predictions about what will happen, and how much of the reactants you would need to make a particular amount of product.

In general, a chemical reaction can be summarised by this equation:

reactant A + reactant B → product C + product D

In every chemical reaction, the amount of matter involved (the mass) is conserved.

The mass of products C and D is the same as the mass of reactants A and B. Although the atoms have been rearranged to form different compounds, no mass has been created or destroyed.

Activity 3: Calculating the amounts of reactants and products

A teacher heats 10 g of copper powder so that it reacts with the oxygen in the air to form copper oxide. She finds the mass of the copper oxide afterwards and it is 12.5 g.

a)   Use the information in the question to write down the word equation for this reaction.

   ....

b)   What mass of oxygen from the air reacted with the copper? ....

The teacher then repeated the experiment with 20 g of copper powder and reacted it completely.

c)   What mass of oxygen would be required now to react with all of the copper? State why you think this.

   ....

   ....

d)   What mass of copper oxide would be produced this time? State why you think this.

   ....

   ....

REMEMBER: CLOSED AND NON-CLOSED SYSTEMS

A closed system describes chemical reactions where all the reactants and products stay together and nothing escapes or is added. A non-closed system describes chemical reactions where reactants and products can enter and leave the reaction.

Examples of non-closed system reactions include ones where gases can enter or leave the place where the reaction happens. This includes combustion, where oxygen is added from the air (like the example of heating copper in Activity 3) and reactions that produce gases that escape, such as making metal salts when you add an acid to a metal.

If the system is non-closed, it is easy to miss out reactants or products in equations or calculations. **Watch out** for gases that escape from containers or join from the air!

Activity 4: Magnesium reacting with oxygen from the air

A class was heating pieces of magnesium ribbon inside a heat-proof container called a crucible. Magnesium is a reactive metal and will react with the oxygen in the air if heated to start the reaction. Once underway, the reaction produces a bright white light.

The teacher told each group to carefully measure the mass of the crucible and magnesium ribbon at the start of the experiment, and the mass of the crucible and product at the end, after the crucible had cooled. He asked them to place a lid on the crucible that would let air in but stop the products from escaping.

a) Is this a closed system or a non-closed system? Explain why.

....

The general equation for the reaction is:

metal + oxygen → metal oxide

b) Complete the word equation for this particular reaction:

.... + ... → ....

c) One group weighed a crucible and magnesium at the start of the experiment and found that it had a mass of 25.5 g. Three learners in the group give different predictions about the mass of the crucible at the end. Which do you agree with? Underline your answer.

Rajiv: 'I think that the mass will be less at the end, as the magnesium is burning away.'

Wing: 'I think that the mass will be the same, as nothing can escape.'

Bella: 'I think that the mass will be more at the end, as oxygen from the air is adding to the magnesium.'

REMEMBER: THE CONSERVATION OF ENERGY

Along with the conservation of mass, the conservation of energy is a really important idea in chemical reactions.

In the chemical reaction that takes place in a Bunsen burner, there is a transfer of energy:

chemical energy → energy released as light + energy released as heat

The chemical energy store is released when molecules of methane react with oxygen in a combustion reaction, as follows:

methane + oxygen → carbon dioxide + water

As combustion is an exothermic reaction, it releases thermal (heat) energy from the reactants as they react. Energy is required to break the bonds in the reactants (methane and oxygen). When these atoms rearrange to make new bonds in the products (carbon dioxide and water), this releases a lot more energy.

The energy that is released from rearranging these bonds is what we see and feel as light and thermal energy. The amount of energy during the reaction can be summarised as follows:

$$\text{released (chemical) energy} = \text{energy released as light during the reaction} + \text{energy released as heat during the reaction}$$

This shows the rule of conservation of energy in action.

Activity 5: Modelling the conservation of energy with water

Some learners are modelling the energy in a chemical reaction. To do this, they have labelled some cups with the energy at each stage, as shown. The amount of water in each cup represents the amount of energy in each part of the transfer.

Before

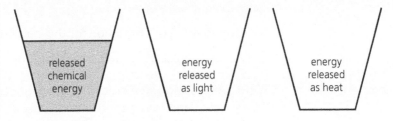

Emma says that the water in the cups after the reaction has taken place should look like this:

Emma

Suki says that the cups should instead look like this:

Suki

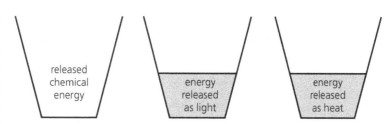

Who do you think has the better answer? Give a reason for your answer.

....

....

....

REMEMBER: CHANGE IN MASS OF REACTANTS

In a non-closed system, reactants and products can enter and leave the reaction container. We can use this idea to help us measure the rate of reaction. If a gas is produced in a reaction, it can escape the container. This means that the reaction container will gradually lose mass to the surroundings. We can calculate the rate of reaction if we know how much the mass has changed in a certain time.

REMEMBER: MEASURING RATES OF REACTION

At the start of this chapter we described rate as how quickly a reaction takes place. We measure this by thinking about how quickly the chemicals change. We look for changes that we can measure; for example, how much the mass of the reactants changes in a certain amount of time, or how long it takes to make a certain amount of product.

We are going to think about two types of change that we can measure:

- the change in the mass of the reactants
- the change in the volume of a (gaseous) product.

Activity 6: Hydrochloric acid and marble chips experiment

A class were measuring the change in mass of the reactants using calcium carbonate (in the form of marble chips) and hydrochloric acid. They used the following equipment to record the loss of mass in two minutes.

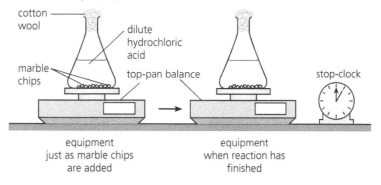

cotton wool

dilute hydrochloric acid

marble chips

top-pan balance

stop-clock

equipment just as marble chips are added

equipment when reaction has finished

a) Complete the following general equation. Use your equation cards to help you.

acid + metal carbonate → ....

....

The mass at the start of the experiment was 157.5 g. At the end it was 155.8 g.

b) Calculate the mass that escaped the container during the experiment.

....

c) Is this a closed or non-closed system? ....

d) Using the word equation for this reaction to help you, describe what could be observed to show that the reaction was taking place.

....

e) Use your answer to part a) to write the word equation for this reaction.

....

....

f) Write a symbol equation for this reaction. Use the formulae for the compounds below to help. Balance the equation if you can.

calcium chloride = $CaCl_2$ hydrochloric acid = HCl
calcium carbonate = $CaCO_3$

....

REMEMBER: CHANGE IN VOLUME OF A PRODUCT

Another way to measure the rate of reaction is to collect the product, measuring the amount produced in a certain time. For reactions that produce gases, we can measure the volume of gas produced in a certain time. To do this, we need to collect the gas in a gas syringe. As the gas is produced, it pushes the plunger in the syringe outwards and the volume produced every minute can be measured.

Activity 7: Hydrochloric acid and magnesium experiment

Some learners were measuring the rate of the reaction between hydrochloric acid and magnesium.

a) Complete the following general equation. Use your equation cards to help you.

acid + metal → .... + ...

The learners carried out the experiment using the equipment shown.

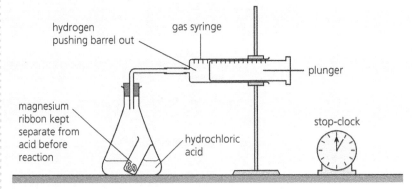

b) Is this a closed or non-closed system? ....

In this section, we will look at three factors that can affect the rate of a reaction:

- concentration
- particle size (not to be confused with the particle model)
- temperature.

c) Using the chemical equation for this reaction to help you, describe what could be observed to show that the reaction was taking place.

....

d) Write the word equation for this reaction. Use your answer to part a) to help you.

....

....

e) Write the symbol equation for this reaction. Use the formulae for the compounds below to help. Balance the equation if you can.

magnesium chloride = $MgCl_2$

....

REMEMBER: CONCENTRATION

The concentration of a solution is a measure of the amount of solute dissolved in a certain amount of solvent. A solution of low concentration has a small amount of solute dissolved in it. A solution of high concentration has a large amount of solute dissolved in it.

If the concentration of a reactant is increased, the rate of reaction increases. If the concentration of one reactant is doubled, the rate of the reaction may be doubled.

A concentrated solution has more particles in it that are available to react than a dilute solution.

This means that increasing the concentration of a solution increases the number of particles, and so increases the number of collisions and the rate of reaction.

For example, a higher concentration of acid would increase the rate of reaction with a metal carbonate. This is because there are more acid molecules to bump into the carbonate molecules and make the reaction occur, releasing carbon dioxide more quickly.

Activity 8: Predicting rate with changing concentration

Some learners are investigating the rate of a reaction between sulfuric acid and exactly the right amount of copper carbonate. They measure the volume of gas produced over time.

The graph below shows their results for a particular concentration of sulfuric acid.

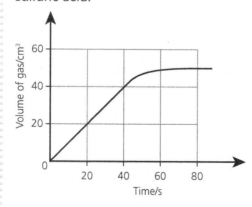

a) Use your knowledge of the reactions of acids with metal carbonates to name the gas that would be produced in the reaction.

....

b) On the graph above, sketch a new line to show what you would expect the results to be if the concentration of sulfuric acid was doubled. (Assume everything else in the experiment stays the same.)

REMEMBER: PARTICLE SIZE

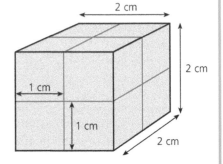

Particle size can affect the rate of a chemical reaction, as it affects the amount of exposed surface area of a reactant.

If you take a large cube and split it into eight smaller cubes, the surface area increases, as shown in the diagram.

The total surface of one large cube:

= number of cubes × number of sides × area of each side

=   1   ×   6   ×   2 cm × 2 cm

= 24 cm²

Total surface of eight small cubes:

= number of cubes × number of sides × area of each side

=   8   ×   6   ×   1 cm × 1 cm

= 48 cm²

The total surface area has doubled by cutting the cube into smaller cubes.

The surface area is important as reactions take place at surfaces, where the particles can collide. The larger the total surface area, the greater the chance of collisions between the particles on the surface and the particles of the reactant in the liquid or gas next to the surface.

This means that increasing the surface area increases the chance of collisions and so increases the rate of reaction.

Activity 9: Custard power explosion risk

Custard powder is made in a factory using cornflour. Cornflour is made by grinding maize seeds into a fine dust, making the particle size smaller.

Maize seeds will burn, but the reaction takes place quite slowly. However, in the factory where maize seeds are ground into flour, there is a risk of explosion. Loose maize dust is therefore extracted from the air by filters.

Use your knowledge of particle size and how it affects the rate of reaction to suggest why fine maize dust might be explosive.

....

....

....

Activity 10: Predicting rate with changing particle size

Some learners are making a salt using zinc granules and hydrochloric acid. The granules are available in three sizes: small, medium and large.

The learners investigate the rate of reaction by looking at the reduction in mass in a given time. They keep the same starting mass for each experiment. They also keep the temperature of the acid the same and use the same concentration of hydrochloric acid.

The results from their experiment are shown in the table.

Particle size	Mass of container at start of experiment/g	Mass of container at end of experiment/g	Mass lost during experiment/g
small	185.5	181.4	
medium	185.5	182.7	
large	185.5	183.9	

a) Calculate the change in mass in each experiment and complete the last column of the table.

b) Describe the relationship between particle size and change in mass.

....

c) Describe the relationship between particle size and the rate of reaction.

....

REMEMBER: TEMPERATURE AND PARTICLE SPEED

The speed at which particles move depends on their temperature. If the temperature is raised, the speed of the particles increases. This means they make harder and more frequent collisions, which are more likely to result in reactions. The rate of reaction therefore increases.

If you place a substance that reacts with water, such as calcium, in warm water, the warm water particles collide more often with the calcium than they would if the water was cold. They also hit the calcium harder, making it more likely to react.

Activity 11: Predicting rate with changing temperature

Some learners are investigating the rate of reaction of copper carbonate with sulfuric acid at different temperatures. The learners use the same amount of copper carbonate each time and the same concentration of acid, but change the temperature of the acid for each experiment. They collect the gas produced and record the time it takes to fill the container.

The learners' results are shown in the list below, but they are jumbled up. Use your understanding of how rate of reaction is affected by temperature to put these results into the table in what you think is the most likely order.

32   63   123   45   89

Temperature of acid (°C)	Time taken to fill the container (s)
20	
30	
40	
50	
60	

REMEMBER: CATALYSTS

Scientists use the word 'catalyst' to mean something that speeds up a reaction. A catalyst is not a reactant or a product and does not react itself.

Examples of catalysts include platinum and rhodium, which are used in vehicles in catalytic converters to turn poisonous carbon monoxide into less harmful carbon dioxide. Iron is a catalyst that is used to speed up the Haber process, which is used to make ammonia for fertilisers.

Activity 12: Catalysts

a) How can catalysts be used to help improve air quality in cities?

....

b) Name **two** examples of catalysts.

....

TAKE A BREAK

Look for examples of fast and slow chemical reactions in your kitchen – remember, in a chemical reaction, a new substance is always made. You might think about different types of cooking and the changes involved. You might also think about unwanted changes like milk turning sour.

Think about how you would speed up or slow down some of these changes. And while you are in the kitchen, grab yourself a drink of water to keep yourself hydrated.

REMEMBER: THE TRIANGLE OF FIRE

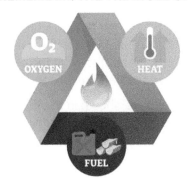

The triangle of fire is a model showing the three features needed for a fire to burn. Removing any one of them can control the fire and put it out.

Activity 13: Extinguishing and preventing fires

In each of the following examples, a fire has been extinguished or prevented in a particular way. State which part of the fire triangle has been removed to extinguish or prevent the fire:

a) Smothering a chip-pan fire with a fire blanket ....

b) Spraying cold water at the base of a fire ....

c) Smothering a fire with a carbon dioxide fire extinguisher ....

d) Removing a magnifying glass that was focusing light rays from the Sun ....

e) Removing a pile of empty cardboard boxes from next to a building ....

REMEMBER: THE PARTICLE MODEL AND RATES OF REACTION

The particle model of matter can be used to explain the factors that affect rates of reactions. Particles take part in reactions when they collide with each other, so any factors that increase the chance of collisions will increase the rate of reaction.

Activity 14: The particle model

Draw the particles in: a) a cool liquid and b) a hot liquid. Indicate the speed of their movement with arrows.

Activity 15: Revision poster

Revisit the poster you made at the start of this chapter in Activity 1. Check that what you have included is both complete and accurate. Is there anything you want to change or add? If so, do this now.

TIPS FOR SUCCESS

Go back over the work that you have done in this chapter, to remind yourself of all the information you have covered. When you are ready, complete this short test.

As you work through it, you can help yourself by:

- reading each question carefully – check you understand the question
- looking for key words to use in your answer
- answering the question in your mind first, before you write it down
- making sure you use correct scientific vocabulary in your answers
- using a piece of spare paper to draft any extended answers first, then when you are happy with it you can write your answer in this book
- checking your answers to make sure that you do not want to make any changes.

Revision test

1 Complete the following table by ticking the correct column to show whether each statement is true or false. [4 marks]

Statement	True	False
Closed systems do not allow chemical reactions to take place.		
Closed systems do not allow reactants to enter or products to leave.		
A gas escaping a reaction would be an example of a non-closed system.		
Air entering a reaction from the surroundings is an example of a non-closed system.		

2 Choose the **false** statement. Circle one answer. [1 mark]
   A Energy is neither created nor destroyed during chemical reactions.
   B If more energy is stored in the bonds at the end of the reaction than at the start, energy will be released to the surroundings.
   C If more energy is stored in the bonds at the start of the reaction than at the end, energy will be released to the surroundings.
   D If more energy is stored in the bonds at the end of the reaction than at the start, energy will be taken in from the surroundings.

3 The fire triangle reminds us of the conditions needed for a combustion reaction to happen. Which of the following is **not** part of the fire triangle? Circle one answer. [1 mark]
   A Oxygen   B Fuel   C Heat   D Matches

4 Write the following word equations as symbol equations. [3 marks]
   a) methane + oxygen → carbon dioxide + water

   ....

   b) hydrochloric acid + calcium carbonate → calcium chloride + water + carbon dioxide

   ....

   c) hydrochloric acid + magnesium → magnesium chloride + hydrogen

   ....

5 In the following reactions, use the idea of conservation of mass to calculate the missing masses. [2 marks]

Reactant 1	Reactant 2	Product 1	Product 2	Product 3 (if applicable)
36.5 g of hydrochloric acid	12 g of magnesium	.... g of magnesium chloride	2 g of hydrogen	
9.8 g of sulfuric acid	10 g of calcium carbonate	13.6 g of calcium sulfate	1.8 g of water	.... g of carbon dioxide

6 Some learners are carrying out an experiment to measure the rate of a **non-closed** reaction between zinc and sulfuric acid.
   a) What is the word equation for the reaction between zinc and sulfuric acid? [1 mark]

   ....

   b) Describe a suitable method for measuring the rate of reaction between zinc and sulfuric acid. [4 marks]

   ....

   ....

   ....

   ....

7   A class is carrying out an experiment to measure the rate of reaction of magnesium carbonate with nitric acid in a **closed system**.

a) What is the word equation for the reaction between magnesium carbonate and nitric acid? [1 mark]

   ....

b) Describe a suitable method for measuring the rate of reaction between magnesium carbonate and nitric acid in a closed system. [4 marks]

   ....

   ....

   ....

   ....

8   Use the particle model to describe and explain what happens to the rate of reaction when you increase the concentration of sulfuric acid in a reaction with magnesium. [3 marks]

   ....

   ....

   ....

   ....

   ....

9   Use the idea of surface area to describe and explain what happens when you increase the size of calcium carbonate (marble) pieces in a reaction with hydrochloric acid. [3 marks]

   ....

   ....

   ....

   ....

   ....

10 Use the particle model to describe what happens when you increase the temperature of hydrochloric acid in a reaction with sodium thiosulfate solution. [3 marks]

   ....

   ....

   ....

   ....

   ....

Chapter 12 Energy

CHAPTER INFORMATION

This chapter will help you to revise your learning about energy. By the end of this chapter you should be able to:

- describe the relationship between thermal energy (heat) and temperature
- describe how thermal energy can be used to compare different substances
- describe how energy is conserved
- describe heat dissipation and how heat is transferred through conduction, convection and radiation
- explain how the structure of metals is linked to conduction
- explain how evaporation cools things down
- give uses of thermal imaging.

REVISION APPROACH: THREE PILES

You are going to begin this chapter by using a revision strategy called 'three piles'. The aim of this approach is to help you decide which scientific words you know and can use.

The three piles are shown in the table.

Know and can use	Not sure	Don't know
I am confident that I know this word and its definition and can use this word correctly.	I have heard of this word; I think I know what it means but I am not sure I am right.	I don't know this word, or I have heard this word but do not know what it means.

The aim as you work through this chapter is to develop your understanding of the key words and move all the words to the 'Know and can use' pile, as you become more confident.

Activity 1: Three piles

a) Make a card for each of the words/phrases in the list below. You only need to write the word on the card for now.

conduction	energy	radiation
conservation of energy	evaporation	temperature
convection	exothermic reaction	thermal conductors
electrons	heat dissipation	thermal energy
endothermic reaction	internal energy	thermal imaging
	kinetic energy	thermal insulators

b) Think carefully about each word and assess how confident you are about it. Then sort the words into the three piles: 'Know and can use', 'Not sure' and 'Don't know'. When you have finished, take a photograph of the piles or copy the words into your notebook.

c) Take the word cards from the 'Know and can use' pile and add a simple illustration on the front of each card. On the back of each card, write as much information as you can linked to the word. This could include a definition, facts or examples. Keep the cards in a safe place, as you will repeat this activity throughout this chapter.

REMEMBER: THERMAL ENERGY AND KINETIC ENERGY

The amount of thermal energy stored in an object is a measure of the **total** amount of kinetic energy of the atoms and molecules in the substance (internal energy). Thermal energy is measured in joules (J). Thermal energy moves around to different stores by processes called conduction, convection, radiation and evaporation.

Thermal energy is not the same as temperature. Temperature is a measure of the **average** kinetic energy of the particles in a substance. Temperature is measured with a thermometer, usually in degrees Celsius (°C).

Activity 2: Energy transfers

Some learners carry out an experiment to investigate the transfer of energy to different substances using the equipment shown.

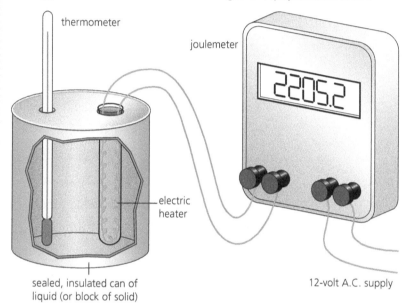

thermometer

joulemeter

electric heater

sealed, insulated can of liquid (or block of solid)

12-volt A.C. supply

They compare the different substances by placing the same mass of each substance in the metal can. The substances are then heated from room temperature (20°C) to a temperature of 50°C. The results are shown in the table.

Substance	Energy reading 1 (joules)	Energy reading 2 (joules)	Energy reading 3 (joules)	Mean energy reading (joules)
water	13 102	13 006	13 207	
vegetable oil	5551	5335	5422	
soil (wet)	8131	8059	8032	
sand	2919	3022	2858	

Use the data collected to answer the following questions.

a) Calculate the mean value for each substance to complete the table. Remember, the mean is the total of the readings divided by the number of readings.

b) Which substance required the largest amount of energy for this temperature change? ....

c) Which substance required the smallest amount of energy for this temperature change? ....

d) Which substance stored the most energy? ....

e) Which substance stored the least energy? ....

f) If the substances were all heated to a higher temperature, would the readings on the joulemeter be higher or lower than in this experiment? Explain your answer.

....

....

Activity 3: Checking learner answers

A learner was asked to complete some sentences about thermal energy, but they have made some mistakes. For each sentence, circle 'yes' or 'no' to show whether it is correct. If it is not correct, write a correction.

a) As a substance is heated, the particles ... *move around less.*

Is it correct? yes / no   Correction ....

....

b) Particles that are moving around more ... *have greater kinetic energy.*

Is it correct? yes / no   Correction ....

....

c) When you measure temperature ... *you use a thermometer.*

Is it correct? yes / no   Correction ....

....

d) When a thermometer is put in a liquid ... *it measures the total energy of the particles in the liquid.*

Is it correct? yes / no   Correction ....

....

> **REMEMBER: CONSERVATION OF ENERGY**
> Like mass, energy cannot be created or destroyed. It is simply transferred from one store to another. This is known as the **conservation of energy**. It often spreads out into multiple stores, which is known as **dissipation**.

Activity 4: Conservation of energy

A teacher is explaining the idea of conservation of energy. They say that sometimes it can be hard to find the energy after a chemical reaction or an energy transfer, because the energy is dissipated (spread out among several stores). To illustrate this to the class, the teacher uses building blocks to give examples of different types of interaction. The piles of blocks represent amounts of energy.

Each example is missing some key pieces of information. Complete the table by adding in the missing energy values and store types.

Example of energy transfer	Energy store at start – amount and type of store	Energy store at end (1) – amount and type of store	Energy store at end (2) – amount and type of store
Switching on a torch	☐☐☐☐☐ Chemical store in the battery	☐ Transferred as light, ending up in a thermal store when light is absorbed	
A kettle cooling down to room temperature	☐☐☐☐☐ ☐☐☐☐☐ Thermal store in the kettle		
An exothermic chemical reaction (e.g. lighting a match)		☐☐☐☐☐ ☐☐☐ Transferred to surroundings, ending up in a thermal store	☐☐ Transferred as light to surroundings, ending up in a thermal store

---

REMEMBER: HEAT DISSIPATION

Thermal energy always moves from a warmer object or place to a cooler object or place. It will continue to transfer until there is no longer a difference in temperature between the two objects or places.

REMEMBER: THERMAL ENERGY MOVES

Thermal energy moves around in substances and from one object to another. It does this by the processes of:

Conduction – when energy is passed by vibrations from particle to particle. This happens best when materials are solids and even better when they are metals, as the 'sea of electrons' in metals passes the energy on between the particles. Some materials (for example, most non-metals) are poor conductors and are called **insulators**.

Convection – when particles in liquids and gases move around within a substance. Warmer areas of liquid/gas are less dense and rise to the top, and the particles carry the energy with them.

Radiation – when infrared radiation radiates from the surface of an object.

Evaporation – when particles at the surface of a liquid have enough energy to leave the liquid as a gas, taking some of the energy with them.

An example that shows all the processes of thermal energy movement in action is a cup of coffee sitting on a desk as it cools down. Fill in the information in each box to describe the main process at each different part of the cup. The name of the first process is included for you.

From the top surface of the liquid
Name of process: ....
Describe what is happening:
....
....
....
....

Inside the liquid
Name of process: Convection
Describe what is happening:
....
....
....
....

From the side surface of the cup
Name of process: ....
Describe what is happening:
....
....
....
....

Through the bottom of the cup
Name of process: ....
Describe what is happening:
....
....
....
....

The following key words have been used so far in this chapter.

conservation of energy	heat dissipation	temperature
electrons	internal energy	thermal energy
energy	kinetic energy	

a) Take these words from your word card pile. If you already knew any of these words and created a card for them, check the information on each card to make sure it is correct. Ask someone else to check those cards as well.

b) Create a card for each word that was in the 'Not sure' and 'Don't know' piles. On the back of each card, write as much information as you can linked to the word. If it helps you to remember the words, draw pictures too.

c) Ask someone to test you on each of the words from this activity. If you are still unsure of any of the words, find a way of learning them so that by the time you finish this chapter, you are confident in using those words.

REMEMBER: THERMAL IMAGING

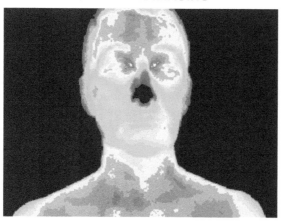

Thermal imaging uses radiation to enable us to see different temperatures in an object or space. The thermal imaging camera receives infrared waves and produces an image with false colour to show hotter areas in red/orange and cooler areas in green/blue. Thermal imaging has many uses, for example:

- showing what is behind clouds of smoke or within rubble, allowing rescue workers to find people trapped in collapsed or burning buildings
- assessing body temperatures of humans and animals
- monitoring volcanic activity by detecting new fissures (cracks) and surface temperatures to help predict future eruptions
- detecting faults in pipelines carrying hot liquids, which is useful for identifying faults in machinery.

Activity 7: 3–2–1

a) When you have read the information above, write down **three** things that you already knew about thermal imaging.

1 ....

2 ....

3 ....

b) Write down **two** things that you did not know, and describe how you will remember them.

1 ....

2 ....

c) Write down **one** thing that you found interesting. Explain why.

1 ....

....

TAKE A BREAK

Chat to a friend for 5 minutes about something completely different from what you are learning. Give your brain this quick break, and then return to your revision.

Activity 8: Apply your understanding

Learners were asked this question:

Explain how thermal energy moves by conduction, convection and radiation.

Using the mark scheme provided, assess the answers below from three different learners.

Includes	Mark
Correctly identify the direction that thermal energy would travel, given information about the temperature of two objects or places.	1
Correctly match the process of conduction, convection, radiation, or evaporation to the energy transfer.	1
Describe the process of conduction, convection or evaporation by using the particle model to describe how particles move during the transfer or describe the process of radiation by using the idea of waves.	2
Connect the ideas together well, so that the explanation is clear and makes sense.	1

Answer A:

In metals, thermal energy transfers mostly via conduction. If you heat one end of a metal rod, the heat travels to the opposite end quite quickly.

This is because the particles in the metal are really close together and some of their electrons form a sea of electrons that pass the energy between the atoms really well.

Mark and reasoning ....

....

....

Answer B:

In water, the energy passes mostly via conduction, although the water is not a very good conductor. The particles are quite close together in a liquid, but not as close as they are in a solid.

Mark and reasoning ....

....

....

Answer C:

A hot surface will pass energy by radiation. Some surfaces are better at radiating than others. A hot object will pass energy to its surroundings. Radiation involves the surface sending particles out in all directions.

Mark and reasoning ....

....

....

Activity 9: Check all words

a) Look back through all the words listed in Activity 1 at the start of this chapter. Take all the words you placed in the 'Not sure' and 'Don't know' piles and add as much information to each card as you can, having learned about them in this chapter.

b) Ask someone to test you on each of the words.

c) Are there any words you are still unsure of or don't know? Share these with your teacher and decide how you will learn these words.

> **TIPS FOR SUCCESS**
>
> Go back over the work you have done in this chapter to remind yourself of all the information you have covered. When you are ready, complete this short test.
>
> As you work through it, you can help yourself by:
>
> - reading each question carefully – check you understand the question
> - looking for key words to use in your answer
> - answering the question in your mind first, before you write it down
> - making sure you use correct scientific vocabulary in your answers
> - using a piece of spare paper to draft any extended answers first, then when you are happy with it you can write your answer in this book
> - checking your answers to make sure that you do not want to make any changes.

Revision test

1 Which of the following statements about thermal energy is correct? Circle the correct answer. [1 mark]

  A Temperature means exactly the same thing as thermal energy.

  B Thermal energy is lower if the object is warmer.

  C The amount of thermal energy stored in an object is a measure of the total amount of kinetic energy of the atoms and molecules in a substance (internal energy).

  D Thermal energy always disappears.

2 Which of the following statements about temperature and thermal energy is correct? Circle the correct answer. [1 mark]

  A Temperature and thermal energy are both measured in joules.

  B Temperature and thermal energy are both measured in degrees Celsius.

  C Temperature is measured in joules; thermal energy is measured in degrees Celsius.

  D Temperature is measured in degrees Celsius; thermal energy is measured in joules.

3 Draw lines to match each sentence opening on the left with the correct ending on the right. [4 marks]

Thermal energy is …	… the average amount of kinetic (movement) energy of the particles in a substance.
Temperature is …	… the total amount of kinetic (movement) energy of all of the particles in an object.
Compared with a glowing metal light filament, a bath of warm water …	… has the higher temperature.
Compared with a bath of warm water, a glowing metal light filament …	… has the larger amount of thermal energy.

4 Draw lines to match each sentence opening on the left with the correct ending on the right. [4 marks]

Conduction …	… explains why warm air rises.
Convection …	… can be detected through infrared imaging.
Radiation …	… occurs when particles leave a surface and take the thermal energy with them, cooling the surface down.
Evaporation …	… happens through solid substances, particularly metals.

5 Give an example of a use of thermal imaging. Describe how different parts of the object being viewed in your example give out different amounts of radiation. [2 marks]

....

....

....

....

6 People usually sweat after exercising, which helps cool them down. Give the name of the process that cools them down and explain how it does this. [2 marks]

....

....

....

....

7 Explain, using an example, what is meant by 'conservation of energy'. [2 marks]

....

....

....

....

8 Explain why the structure of metals makes them very good conductors of thermal energy. [2 marks]

....

....

....

....

9 Describe what is meant by a thermal insulator and give an example. [2 marks]

....

....

....

....

Chapter 13 Waves

CHAPTER INFORMATION

This chapter will help you to revise your learning about waves in the context of sound. By the end of this chapter you should be able to:

- use the particle model to explain how sound waves are formed when particles vibrate
- identify the main features of a waveform
- explain how amplitude is linked to loudness
- explain how pitch is linked to frequency
- describe what is meant by the Doppler effect
- explain what happens when sound waves interact
- explain how to model sound waves.

REVISION APPROACH: HEXAGONS CONCEPT MAP

To help you remember the key words of this topic, you are going to make key word cards. However, for this topic your key word cards will be in the shape of hexagons. As hexagons tessellate, you can place the hexagon cards against each other to show links between the words and help your understanding.

Activity 1: Hexagon key word cards

Here is a list of the words that you need to learn and remember about sound waves:

acoustics	destructive	pressure
air particles	frequency	rarefaction
amplitude	hertz (Hz)	sound energy
compression	noise pollution	sound wave
constructive	particles	waveform
decibels	pitch	wavelength

a) Write each word on the front of a hexagon-shaped card. Add any images that will help you remember what the word means.

b) On the back of each card, write a definition. Add as many other ideas as you can that link to the key word on the front. Do not worry if you cannot do this for all the cards yet.

c) Group the hexagons according to the information on each card. Now arrange the hexagons on a sheet of paper so that they fit together. Remember, the information on each hexagon must link to the hexagons around it. As you work through this chapter, add any new cards to your arrangement, even if it means moving some of the cards you have already put down.

d) Take a photograph of the way you have laid out your hexagons. In one of the later activities, you will be asked to look back at the arrangement of hexagons that you made.

Activity 2: What causes sound?

Imagine you have an object vibrating back and forth in air so that it produces a sound.

Order the following sentences so that they form an explanation of how the sound is formed. The first and last statements have already been listed in the sequence to help you.

A   The object that is vibrating moves towards the air particles nearest its surface.

B   The object then moves in the opposite direction, away from these air particles.

C   The particles in the air nearest the object are then less tightly packed than they would otherwise be. This causes a low-pressure zone called a rarefaction.

D   As the object moves back in the first direction, it causes another compression.

E   The object pushes these air particles together, so they are more tightly packed than usual. This causes a high-pressure zone called a compression.

F   The continuing back and forth motion causes a series of pressure variations, which can travel through the air as sound.

The correct sequence is: A, ..., ..., ..., ..., F.

> **REMEMBER: SOUND NEEDS PARTICLES TO TRAVEL**
> Sound can only travel through a medium that contains particles. The particles pass on the vibrations as waves of compressions and rarefactions. The more closely the particles are packed, the faster the sound will travel, as the particles do not have to travel as far before hitting another particle and passing the vibration on. If there are no particles, sound vibrations cannot travel as there is no medium to pass them on.

Activity 3: Sound travelling

Three friends are sitting in a coffee shop, discussing the music that they can hear from the radio. They notice that they cannot hear the music as well from where they are sitting as they could from where they queued up to order their coffee.

> **Bailey:** I think the sound is quieter here because there is a solid partition in between us and the radio and sound can't travel through solids at all.

> **Muhammed:** I think it's quieter because the sound particles aren't reaching us here.

> **Cho:** I think one of the reasons the sound is quieter is because we are sitting quite a distance away from the radio.

Only one of their statements is fully correct. Which do you think it is? Give a reason for your answer.

....

....

> **REMEMBER: SOUNDS ARE CAUSED BY VIBRATIONS**
> You might remember that sounds are caused when an object vibrates. There are many different types of object (or parts of objects) that can vibrate and produce sounds.

Complete the labels for the compression and rarefaction diagram below. Use the key words from the list to help you.

low pressure   high pressure   compression

rarefaction   vibrating object   direction of travel of sound

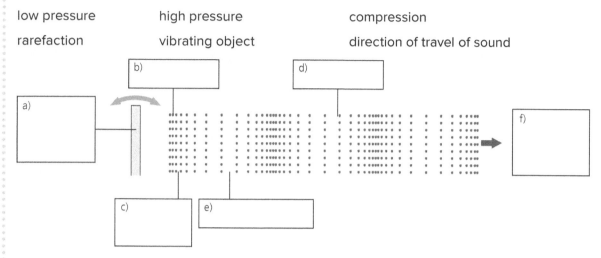

Sound will travel through three of the media shown in the table, but will travel at different speeds. One of the media will not allow sound to travel at all.

Complete the table by adding in the speed of sound that you think matches each medium from the list below. It might help you to think of the particle arrangements in solids, liquids and gases.

2840 m/s   1560 m/s   330 m/s   no sound

Medium	Gas, such as air	Solid, such as glass	A vacuum, such as outer space	Liquid, such as seawater
Speed of sound				

REMEMBER: DIFFERENT TYPES OF SOUNDS

The speed at which an object vibrates will alter the number of waves produced every second, which we hear as a change in the pitch of the sound. Objects that vibrate more quickly will give a higher pitched sound.

For example, a long ruler that has been 'twanged' will vibrate slowly and give a lower pitched sound.

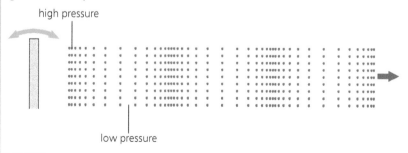

A short ruler that has been 'twanged' will vibrate much faster and give a higher pitched sound.

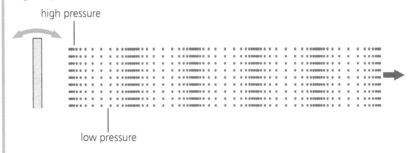

The loudness of a sound depends on the contrast between the areas of high and low pressure.

A long ruler that has been 'twanged' gently will vibrate slowly and give a low-pitched sound. A long ruler that has been 'twanged' harder will still vibrate slowly to give a low-pitched sound, but the contrast between the high and low pressure areas will be greater, producing a louder sound.

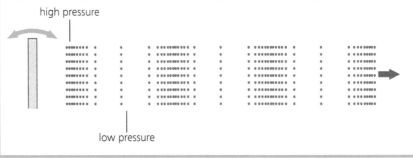

REMEMBER: SHOWING SOUND WAVES

When we capture sounds with a microphone, we can use an oscilloscope to give a picture of the sound. It will show a wave shape on the screen called a waveform. This is a type of graph called a displacement–time graph, showing the distance a particle travels in one direction (position Y) and the distance it travels in the other direction (position Z) over time (shown by the line X). The key features of a waveform are shown in the diagram.

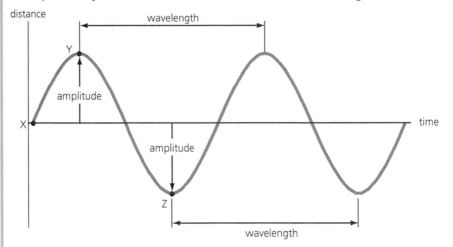

The louder a sound is, the larger the amplitude will be on the waveform. The higher the pitch of a sound is, the shorter the wavelengths will be on the waveform.

Activity 6: Matching waveforms

The diagram shows the oscilloscope trace for a particular sound.

Below is a list of descriptions of some other sounds:

higher pitched sound   louder sound   quieter sound

Match each of these descriptions to the correct oscilloscope trace diagram by writing the description underneath.

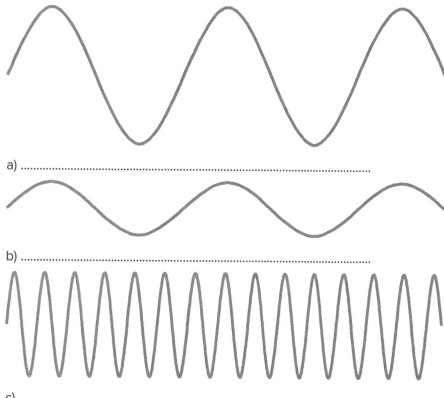

a) ....

b) ....

c) ....

REMEMBER: THE DOPPLER EFFECT

The Doppler effect describes how sounds change when objects that are making a sound are moving towards or away from us. Sounds tend to increase in pitch as they move towards us. Sounds tend to decrease in pitch as they move away from us. The faster they are moving, the bigger the size of the change in pitch.

Activity 7: Police sirens

Zoe and Xander are walking along the street. They notice a police car travelling past at high speed. They hear the car's siren as the car approaches, passes and zooms away.

Draw lines to match the sounds to the descriptions.

As the police car approaches ...	... the sound is a normal pitch.
As the police car is nearest them ...	... the pitch is lower than normal.
As the police car travels away ...	... the pitch is higher than normal.

Here are some key facts that you need to remember about waves and sound:

- Sound is measured in decibels (dB).
- Sounds of 85 dB and above are damaging to your hearing over time.
- Loud noises above 120 dB can cause immediate harm to your ears.

A person's hearing can be damaged by what they do. Some sounds can damage hearing because of how close a person is to a loud sound, such as a balloon popping. Hearing can also be damaged if a person is exposed to a sound for a long time; for example, working in a noisy factory every day, or frequently listening to loud music through headphones.

Sound	Loudness in dB
jet engine taking off	130
music concert	120
ambulance siren	120
vacuum cleaner	70
normal speech	55
whisper	20

REMEMBER: MODELLING SOUND WAVES

Sound waves can be modelled using a large spring, also known as a slinky spring. You can move your hand back and forth to show how the particles collide with one another in sound waves. The sound in this model would travel from left (with the hand representing the source of the sound) to right.

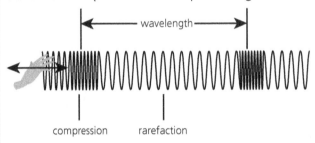

Slinky springs can also be used in a different way by using a side-to-side motion. While this is a good model for other waves, like those on water, **it is not how the compressions and rarefactions work in sound**.

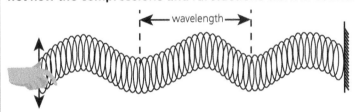

Activity 8: Throwing stones in a pond

Will and Femi are throwing stones in a pond. They notice that the ripples on the surface of the water look like the ripple tank that their teacher had shown them in class. Will says that he is going to try to demonstrate what different sounds would look like by throwing stones into the pond in different ways.

Draw lines to match the following descriptions of the stone-throwing to the sounds that Will was trying to demonstrate.

Loud sound	A few stones, thrown in slowly one after the other
Quiet sound	A large stone
A high frequency	A small stone
A low frequency	A few stones, thrown in quickly one after the other

REMEMBER: WHEN SOUND WAVES INTERACT, THEY DO NOT ALWAYS ADD UP THE WAY YOU WOULD EXPECT ...

Because sounds are waves, putting them together is not always simple. Sometimes they add up to make very loud sounds. Sometimes they cancel each other out. If the peaks of the waves line up, you will get constructive interference. If the peaks of one wave line up with the troughs of another, you will get destructive interference. Remember, constructive waves build up – they produce waves of greater amplitude. Destructive waves destroy – they produce waves of smaller amplitude.

Activity 9 will test your knowledge of constructive and destructive interference to consider why adding two sound waves does not always make a louder sound.

Activity 9: Cancelling out sound?

Aarti is a sound engineer working for a manufacturer of car sound systems. She is talking about the latest type of noise-cancelling speakers that her company has been fitting to executive cars. Aarti claims that the speakers play back noise to cancel out the noise of the engine.

Geoff is confused by her explanation. He thinks that playing more sounds will just add more noise to the car, making it louder. He thinks the noise-cancelling systems would be a waste of money.

What should Aarti say to Geoff to help him understand how the system is designed to work? Use your knowledge of **constructive interference** and **destructive interference** to help.

....

....

....

....

Now that you have completed this chapter, lay out the hexagons you used in Activity 1 again, making links between the cards. Compare this to the way you laid them out the first time. Are you happy with the original connections? Have you made new or stronger connections now? What does this tell you about your learning in this chapter?

TIPS FOR SUCCESS

Go back over the work you have done in this chapter to remind yourself of all the information you have covered. When you are ready, complete this short test.

As you work through it, you can help yourself by:

● reading each question carefully – check you understand the question
● looking for key words to use in your answer
● answering the question in your mind first, before you write it down
● making sure you use correct scientific vocabulary in your answers
● using a piece of spare paper to draft any extended answers first, then when you are happy with it you can write your answer in this book
● checking your answers to make sure that you do not want to make any changes.

Revision test

1 Complete the following table by ticking the correct column to show whether each statement is true or false. [5 marks]

Statement	True	False
Sound needs to travel through a medium containing particles.		
Sound travels fastest in air.		
The higher the amplitude, the louder the sound.		
The shorter the wavelength, the lower the pitch.		
The loudness of a sound is measured in decibels (dB).		

2 Which of the following does **not** cause hearing loss? Circle one answer. [1 mark]
   A Sounds that are very loud
   B Loud sounds that are sustained
   C Wearing ear defenders in a noisy workplace

3 Sounds can be added together in a constructive or destructive way when they interfere with one another. Which of the statements below correctly describes what will happen? Circle the correct answer. [1 mark]
   A Sounds that add via constructive interference will end up louder than either sound alone.
   B Sounds that add via destructive interference will end up louder than either sound alone.
   C Sounds that add via constructive or destructive interference will always change the frequency of both sounds.

4 Draw lines to match the following words with the correct descriptions. [4 marks]

compressions	low pitch
rarefactions	loud sound
high amplitude	high pressure
low frequency	low pressure

5 Match the labels to the spaces on the diagram. [4 marks]

wavelength   amplitude   wavelength   amplitude

6 Explain how a sound wave travels through air to reach someone who is listening. [3 marks]

....

....

....

7 Describe the differences and similarities between the waveforms in each pair of pictures.
Use some of the following words to help you:

higher   lower   same   amplitude
louder   quieter   frequency   wavelength

a) [4 marks]

A

B

....

....

....

....

b) [4 marks]

A

B

....
....
....
....

8 Describe what happens in the Doppler effect. [2 marks]

....
....
....
....
....

9 A teacher shows three actions using a slinky spring to model sound waves. However, there
   are three mistakes in what the teacher shows and describes. Identify each mistake and
   suggest a correction. [3 marks]
   – They move their hand from side to side as shown.

   – They change how often they move their hand from side to side, saying that this will change
     the amplitude.
   – They change how far they move their hand from side to side, saying this will show the
     wavelength.

   Mistake 1 ....
   ....
   Mistake 2 ....
   ....
   Mistake 3 ....
   ....

Chapter 14 Electrical circuits

CHAPTER INFORMATION

This section will help you to revise your learning about electricity. By the end of this chapter you should be able to:

- state the differences between series circuits and parallel circuits
- measure current and voltage in series circuits and parallel circuits
- describe what happens to current and voltage when cells and lamps are added to series and parallel circuits
- describe how resistance affects current
- calculate resistance
- draw series and parallel circuit diagrams that include fixed and variable resistors and buzzers
- make and compare circuits that contain fixed and variable resistors and buzzers.

REVISION APPROACH: ELECTRICITY CONCEPT MAP

In this chapter, you will use a concept map revision strategy. A concept map is a useful way to help you remember key words and ideas in a topic and to check that you have understood how ideas link together.

Activity 1: Key word cards and concept map

a) Look at the list of words below.

ammeter	electrical conductor	resistor
amp	electrical insulator	series circuit
battery	lamp	variable resistor
buzzer	parallel circuit	voltage
cell	resistance	voltmeter
current	resistance = $\dfrac{\text{voltage}}{\text{current}}$	

a) Make word cards for all the words that you recognise. Do not worry if you do not know all the words yet. You can make cards for any words you do not know when you meet them as you go through this chapter. Write each word on the front of a card and add any images that help you remember what the word means.

b) On the back of each card, write as much information as you can linked to the word. This could include a definition, facts or examples.

c) Try to sort the cards you know into groups; for example, names of electrical components.

d) Lay out the word cards you have made on a piece of plain paper.

e) Now draw lines on the paper to connect as many cards as you can. On the connecting lines, write the reason why you have linked the cards. There are lots of different ways that you can link the cards; you just need to be able to explain your reasoning for the links you make.

f) When you have finished, take a photograph of your concept map and collect and keep your word cards safe. You will need them again at the end of this chapter.

REMEMBER: SIMPLE ELECTRICAL CIRCUITS

Electrical circuits can be very complicated, but they all work in the same way. They have a power source connected by wires to different components, such as lamps. We call the power source a cell. When two or more cells are joined together, they form a battery. Circuits are drawn using circuit diagrams consisting of straight lines and symbols representing components.

Activity 2: Circuit symbols

Look at the circuit symbols in the table. Try to memorise as many as you can. Cover the middle column and draw the symbol for each component from memory in the right-hand column.

cell	⊣⊢	
lamp	⊗	
ammeter	Ⓐ	
voltmeter	Ⓥ	
resistor	▭	
buzzer	⏛	

REMEMBER: SERIES AND PARALLEL CIRCUITS

In a series circuit, components are connected in a single loop. If the circuit is broken (for example, if one component does not work) then the rest of the circuit will not work.

In a parallel circuit, components are connected in separate loops and the current follows different paths. Unlike a series circuit, if one component does not work, the others will if they are in different loops. All parallel circuits have multiple loops.

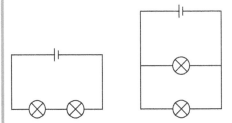

A series circuit (left) and parallel circuit (right)

Activity 3: Broken circuits

In the circuits in the diagram, the lamps circled in red are broken.

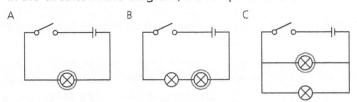

What would happen in each case when the switch was closed? Explain why for each case.

Circuit A: ....

....

Circuit B: ....

....

Circuit C: ....

....

REMEMBER: MEASURING CURRENT

The current flowing in a circuit is measured in amperes (amps), which is abbreviated to A. An ammeter is used to measure current in a circuit. To measure the current flowing through any component, the ammeter must be placed in series with it. Current is the same at all points of a series circuit, so it does not matter where the ammeter is placed in the circuit.

In a parallel circuit, the current will divide where loops in the circuit meet. If there are two loops, the current divides into two, three loops will mean it divides into three, and so on. If the components on each loop are equal, the current will divide between the loops equally. Even if the branches are unequal, the current along each branch will always be the same as the current that goes into the branch.

Activity 4: Measuring current

a) Which of the following circuits shows the correct way to measure current? Circle the correct answer(s).

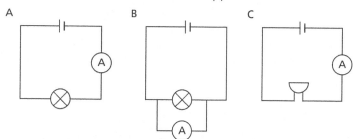

b) The circuit below has two identical buzzers. Calculate the missing ammeter readings.

Ammeter	Current/amps
A_1	1.0
A_2	
A_3	
A_4	

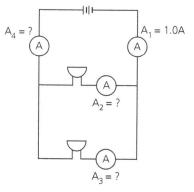

c) Complete the following sentences by circling the correct words.

No matter where you place the ammeter in a loop, the current reading is the same / different. This is because the current is the same / different at all points in a loop.

REMEMBER: MEASURING VOLTAGE

Voltage is the name for the 'push' that causes charges to flow in a circuit. It is linked to the amount of energy that the cell or battery provides to the components in the circuit. The higher the voltage, the greater the energy it can supply. When lamps are added to a series circuit, the voltage divides across the lamps. The more lamps there are in the circuit, the smaller the share of voltage each one receives, making them dimmer.

However, in a parallel circuit, adding a new lamp in a new loop will not change the brightness of the lamps. This is because the energy transferred to the component in each loop is the same as the energy transferred by the battery.

On the side of a cell (battery), you will see the letter 'V' with a number (for example, 1.5 V or 3 V) which indicates the voltage. In a circuit, it is possible to change the voltage supplied by adding cells to make a larger battery. To calculate the overall voltage of the battery, you simply add the voltages of the cells together. For example, a battery made up of two 1.5 V cells has a total voltage of 3 V. It is important to add the cells in series with one another, so that positive terminals connect with negative.

Voltage in a circuit can be measured using a voltmeter. Unlike the ammeter, the voltmeter is arranged in parallel with the components in a circuit.

Activity 5: Measuring voltage

Which of the following circuits shows the correct way to measure voltage? Circle the correct answer(s).

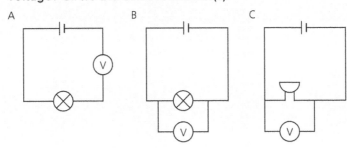

Activity 6: Making a battery from cells

Two groups of learners are putting cells together to make a battery. Diagrams of each group's battery are shown below.

Group A   **Group B**

When the groups connect their batteries to a circuit containing lamps, Group B notice that their lamps are not as bright as those of Group A.

a) What mistake have Group B made when connecting their cells?

....

b) Why do you think their lamps are still working, just not as brightly? (Clue: look at the number of cells in each direction.)

....

....

....

Group A check the voltage of their battery with a voltmeter. The reading is 4.5 V.

c) What do you think the reading would be for the battery of Group B? Circle the correct answer.

    0 V   1.5 V   3.0 V   4.5 V

d) Give a reason for your answer.

....

....

Activity 7: Tweet it

Write a tweet to describe the difference between current and voltage. Remember, tweets have a maximum of only 280 characters.

You might choose to use some of the following words:

current   voltage   charge   flow   energy   loop   circuit   cell/battery

....

....

....

TAKE A BREAK
Stop revising and have a healthy snack, such as fresh fruit, carrot or pepper sticks. Bananas help regulate your blood sugar level, and dried fruits and nuts can help keep your stress levels down. Try to avoid sugary drinks and snacks!

Activity 8: Recalling information

a) Describe what will happen to the brightness of lamps in a circuit as more lamps are added in parallel.

....

b) Give one advantage (not related to brightness) of wiring lamps in parallel.

....

REMEMBER: RESISTANCE

All components in electric circuits have a level of resistance. Resistance tells you how difficult it is for electric current to flow through the components. It is measured in ohms, which have the symbol Ω.

Materials with a very high resistance are called insulators. Examples include plastic and glass. Materials with a very low resistance are called conductors. They are often used in circuits. The best conductors (for example, copper) are used to make wires, as this helps electricity to flow around the circuit.

Resistance is calculated using the following equation:

resistance $(\Omega) = \dfrac{\text{voltage (V)}}{\text{current (A)}}$ $R = \dfrac{V}{I}$

For example: ammeter reading = 0.5 A; voltmeter reading = 1.5 V

resistance $= \dfrac{1.5 \text{ V}}{0.5 \text{ A}} = 3.0 \ \Omega$

TIPS FOR SUCCESS
Remember, when calculating anything, including resistance, you must **show your working**. This includes writing down the formula that you use and substituting in the values for voltage and current carefully to make sure they are correct. It is also helpful to write your numbers clearly, and make your final answer clear, perhaps by underlining it. Finally, don't forget your units.

Activity 9: Calculating resistance

Two circuits have been built by different groups. The groups were asked to calculate the resistance of the component in their circuit.

Look at the circuit each group has built and the measurements they have taken. In each case, work out whether their answer is correct. If it is wrong, try to explain their mistake.

Group A

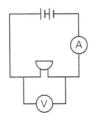

Reading on ammeter = 0.6 A

Reading on voltmeter = 3.0 V

$$\text{Resistance} = \frac{\text{voltage}}{\text{current}}$$

$$= \frac{3.0 \text{ V}}{0.6 \text{ A}}$$

$$= 5.0 \ \Omega$$

Is it correct? If not, why not?

....

....

Group B

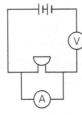

Reading on ammeter = 0.01 A

Reading on voltmeter = 3.0 V

$$\text{Resistance} = \frac{\text{voltage}}{\text{current}}$$

$$= \frac{3.0 \text{ V}}{0.01 \text{ A}}$$

$$= 300 \ \Omega$$

Is it correct? If not, why not?

....

....

REMEMBER: FIXED AND VARIABLE RESISTORS

Sometimes we place components with resistance in a circuit to help control the flow of electric current. These components are called resistors. If the resistance does not change, it is called a fixed resistor. Fixed resistors are often used as protective components in circuits, to control the current flowing through another component and the energy that is transferred by it in order to prevent the component being damaged.

The symbol for a fixed resistor is:

Components that allow the resistance to change are called variable resistors. These change the flow of electric current, so are useful for control devices such as a dimmer switch for a light (where the current to the lamp is increased or decreased to make it brighter or dimmer).

The symbol for a variable resistor is:

Activity 10: Which type of resistor?

Read the descriptions of the circuits below and work out whether the resistor in them is fixed or variable. Circle the correct answers.

Circuit A: The brightness of the lamp in a circuit does not change. **fixed / variable**

Circuit B: The loudness of a buzzer increases or decreases if the resistor is adjusted. **fixed / variable**

Circuit C: The brightness of a lamp increases or decreases if the resistor is adjusted. **fixed / variable**

Activity 11: Revisit your concept map

a) Check that you have made word cards for all the words in the list in Activity 1. If not, make cards for them now.

b) As in Activity 1, use all the cards to create a concept map. Sort the cards into groups. Lay out all the cards on plain paper in groups you can link.

c) When you are happy with the layout, glue your cards onto the paper and draw lines to link them together, writing what the links are. Write the title of each group on the paper.

d) Draw lines to link different groups of cards to help deepen your connections.

e) If there are any words you are still unsure of, ask your teacher to help you with them.

TIPS FOR SUCCESS

Go back over the work you have done in this chapter to remind yourself of all the information you have covered. When you are ready, complete this short test.

As you work through it, you can help yourself by:

- reading each question carefully – check you understand the question
- looking for key words to use in your answer
- answering the question in your mind first, before you write it down
- making sure you use correct scientific vocabulary in your answers
- using a piece of spare paper to draft any extended answers first, then when you are happy with it you can write your answer in this book
- checking your answers to make sure that you do not want to make any changes.

Revision test

1 Which of the following is true? Circle the correct answer.   [1 mark]
   A A variable resistor, adjusted to have a higher resistance, will have a lower current passing through it.
   B A variable resistor, adjusted to have a higher resistance, will have a higher current passing through it.
   C The current through a variable resistor will always stay the same, even if you adjust the resistance.
   D The resistance of a variable resistor will always be zero.

2 Which of the following is true? Circle the correct answer.   [1 mark]
   A Adding more cells to a circuit in series will increase the voltage it supplies.
   B Adding more cells to a circuit in parallel will increase the voltage it supplies.
   C If more lamps are added to a circuit in series with other lamps, they shine just as brightly as a single lamp on its own.
   D If more lamps are added to a circuit in parallel with other lamps, the lamps are dimmer than a single lamp on its own.

3 Which of the following is false? Circle one answer.   [1 mark]
   A The loudness of a buzzer can be controlled with a variable resistor placed correctly in the circuit.
   B The brightness of a lamp can be controlled by a fixed resistor placed correctly in the circuit.
   C The brightness of a lamp will not change if it is in a circuit with a fixed resistor.
   D The resistance of a fixed resistor can be calculated by dividing the voltage across it by the current through it.

4 Complete the following table by ticking the correct column to show whether each statement is true or false.   [5 marks]

Statement	True	False
Voltage is measured in amperes.		
Electric current is measured with an ammeter.		
A series circuit has multiple loops in it.		
Adding more lamps to a circuit always means that they glow as brightly as they would on their own.		
Adding more cells to a battery will always increase the voltage of the battery, as long as they are connected correctly in series.		

5 Draw lines to match each component with the correct description.   [5 marks]

lamp	Used to measure the current in a circuit
ammeter	Always placed in parallel with a component
current	The unit of electric current
voltmeter	The flow of charges around an electric circuit
ampere	Glows when current passes through

6 A teacher provides the following instructions for a circuit.
   – The circuit has a battery made up of two cells.
   – There are two branches to this circuit: the first branch has a buzzer and the second has a lamp.
   – There is a switch before the branch in the circuit.
   – There are three voltmeters in the circuit to measure the voltage across the battery, across the lamp and across the buzzer.

a) Draw the circuit diagram for this circuit in the space below.   [2 marks]

b) Describe what will happen when the switch is closed.   [1 mark]

....

c) If the voltmeter across the battery reads 2.8 V, what would you expect the readings to be on the voltmeter:

   i) across the buzzer? ....

   ii) across the lamp? ....   [2 marks]

d) If the switch is now opened, what would happen to the voltmeter readings?
   Circle the correct answers.   [3 marks]
   i)   The reading on the voltmeter across the battery increases / decreases / stays the same.
   ii) The reading on the voltmeter across the buzzer increases / decreases / stays the same.
   iii) The reading on the voltmeter across the lamp increases / decreases / stays the same.

e) Give a reason for your answers to d).   [1 mark]

....

....

....

7 Fill in the blanks in each statement using the words in the list.   [5 marks]
   series   the same   parallel   lower   divided

a) If the resistance of a component increases, the electric current through the component will
   be ....

b) Ammeters are always connected in .... with the component that you are measuring the current through.

c) Voltmeters are always connected in .... with the component that you are measuring the voltage across.

d) The current in a series circuit is always .... through each component.

e) The current in a parallel circuit is always .... between the branches.

8 A teacher builds the circuit in the diagram. The lamps are all of the same type. Everything in the circuit is working as it should be.

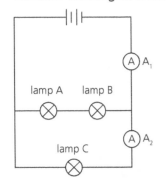

They take the following current readings:

Ammeter	Current/A
A_1	1.5
A_2	1.0

The teacher accidentally knocks lamp B, which falls over and the filament breaks. This stops lamp B from working. None of the other components are damaged.

a) Describe what would happen to the other lamps in the circuit and explain why.
   Use these words:
   current   complete   incomplete   electrical energy   [2 marks]

   ....

   ....

   ....

   ....

   ....

The teacher then replaces all of the lamps with buzzers. The resistance of each of the buzzers is twice as high as the lamps that they have replaced.

b) Predict what the readings will now be on the ammeters in the circuit.   [2 marks]

Ammeter	Current/A
A_1	
A_2	

c) Give a reason for your answer.   [1 mark]

   ....

   ....

Chapter 15 Planet Earth

REVISION APPROACH: PLANET EARTH KEY WORD CONCEPT MAP

You are going to use a revision strategy called concept mapping to help you learn. A concept map is a useful way to help you remember key words and ideas in a topic and check that you have understood how these ideas link together.

Activity 1: Key word cards and concept map

Below is a list of words from this chapter. Do not worry if you do not yet know all the different key words listed. This is the list of words that you need to learn and remember about planet Earth by **the end** of revising this chapter.

convection currents	magnetic polarity	S waves
crust	magnetism	sedimentary
earthquakes	mantle	seismic waves
fossils	metamorphic	tectonic plate
igneous	mid-Atlantic ridge	volcanoes
inner core	outer core	
magma	P waves	

a) Make word cards for all the words that you recognise. Write each word on the front of a card and add any images that help you remember what the word means.

b) On the back of each card, write as much information as you can about the word. This could include a definition, facts or examples.

c) Try to sort the cards you know into groups; for example, types of rocks.

d) Lay out the word cards you have made on a sheet of plain paper.

e) Now draw lines to join as many of the cards as you can on the sheet, and on the connecting lines write the reason why you have linked the cards. For example, you could draw a line between the words 'igneous' and 'magma', and write on the line, 'igneous rocks are formed when magma cools down and solidifies'.

f) The more lines you draw, the more links you can add that show your understanding. There are lots of different ways that you can link the word cards; the key is that you can explain the reasons why you think the connected cards are linked.

g) When you have finished, take a photograph of your concept map and collect and keep your word cards safe. You will need them again at the end of this chapter.

REMEMBER: TECTONIC PLATES

Scientists have gathered evidence over many years showing that the Earth's solid outer crust is separated into plates that can move over the molten upper portion of the mantle. These large plates are made of massive, irregularly shaped slabs of solid rock. They are called tectonic plates and they form a large portion of the Earth's crust.

Tectonic plates have been drifting and moving about on the surface of the Earth like slow-moving bumper cars, which move together and then separate, for millions of years.

Activity 2: Match evidence for tectonic plates

Scientists have undertaken different activities to collect evidence for the existence of tectonic plates.

a) Draw a line to match each activity with the evidence gathered.

1 Mapping of the Earth's landmass	**A** The mid-Atlantic ridge is part of a range of mountains that runs from pole to pole. This mountain ridge formed as tectonic plates moved apart, releasing magma. This magma was then cooled by the water surrounding it. Mountains are also measured and observed to be growing larger.
2 Analysing igneous, metamorphic and sedimentary rock distributions	**B** Similar plant and animal remains have been found at different locations on the planet. Evidence of animals that only lived in freshwater have been found in South America and Africa, yet these continents are separated by vast volumes of seawater. There are also examples of the same types of plants that have been found in South America, South Africa, Australia and Antarctica.
3 Mapping of ocean floors	**C** The locations where earthquakes happen, as well as where volcanoes exist, show distribution patterns. These patterns indicate that earthquakes and volcanoes exist along edges of tectonic plates.
4 Analysing fossil distributions	**D** The complementary 'jigsaw appearance' of continental coastlines suggests that continents have previously been connected.
5 Mapping of volcanoes and earthquakes	**E** As magma at the edge of a tectonic plate escapes and cools under the ocean, the minerals in the rock line up with the Earth's magnetic field. These cooled rocks form magnetic stripes on either side of the gap. Over millions of years, the magnetic field of the Earth changes direction and this changes the alignment of the next magnetic stripe. Over time, the magnetic stripes form patterns as they move away from each other.
6 Monitoring magnetic polarity of the ocean floor	**F** Similarities have been found in rocks around the world. On the west coast of Africa and the eastern coast of South America, the same type of rock has been found that formed at the same time over two billion years ago.

b) Which of the pieces of evidence above help to explain the theory that all of the land mass on the Earth was once a supercontinent? Explain your reasons.

....

....

....

c) Which of the pieces of evidence above show that there is continuous activity occurring on Earth at the boundaries of tectonic plates? Explain your reasons.

....

....

....

REVISION APPROACH: RICH PICTURE

A rich picture is a way of showing an idea, information or a process by using pictures, diagrams and individual words, phrases and colour-coding. A rich picture is different from an infographic because it does not have to use graphs, charts and numbers. Using a rich picture can make it easier to show what you know than, for example, writing sentences or paragraphs, especially if you are someone who learns and remembers pictures more easily.

Activity 3: Rich picture

Use the steps below to create a rich picture about the different pieces of evidence that explain the existence of tectonic plates.

a) At the centre of your rich picture, draw a large map of the outline of the land on the Earth. The picture below will help you with this.

b) Here is a list of six activities that scientists have undertaken:
- mapping of the Earth's landmass
- analysing igneous, metamorphic and sedimentary rock distributions
- mapping of ocean floors
- analysing fossil distributions
- mapping of volcanoes and earthquakes
- magnetic polarity of the ocean floor.

c) Add six information boxes spread around the outside of your drawing, one for each of the activities and pieces of evidence you matched in Activity 2. Write notes in each box about the evidence the scientists gathered and how it explains the existence of tectonic plates. You may need to do some research to help you with this.

d) Add an image to represent each activity – you will use these images elsewhere on the poster. For example, for mapping fossils you might add an image of a fossil. Whenever you mention mapping fossils on the poster, you can add the same image to symbolise this.

e) Draw arrows from each information box to the places on Earth where scientists gathered this evidence and add the image you have used in your key to show where this happened on Earth.

f) Now use your rich picture to help you revise. See if you can name from memory all six boxes, the key information and the locations on Earth where evidence was collected. Write as much as you can on a new piece of paper, and keep trying this until you have memorised the whole picture. You can use your rich picture for at-a-glance information over the coming days too.

Activity 4: Earth's layers

A Stage 9 learner is trying to learn the different layers of the Earth. They have produced a labelled diagram as shown.

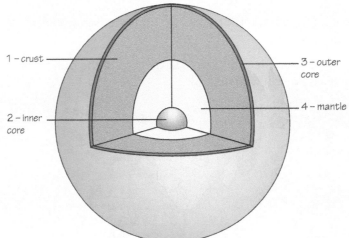

1 – crust

2 – inner core

3 – outer core

4 – mantle

a) Can you spot any mistakes the learner has made with their labelling?

....

....

b) Make up a mnemonic to help you remember the layers in order.

....

REMEMBER: SEISMIC WAVES

We have not managed to dig to great depths in the crust of the Earth. To help them discover what lies beneath, scientists have gathered and analysed data from earthquakes using instruments called seismometers.

When an earthquake occurs, it sets off two kinds of seismic waves that travel down into the Earth. These two different seismic waves are known as primary waves (P waves) and secondary waves (S waves). Seismometers measure these waves.

Primary waves travel through the liquid outer core and the solid inner core. When they pass from one to the other, they change their path (refraction). They travel faster than S waves. Secondary waves cannot travel through the liquid outer core and are reflected from the liquid surface. They travel more slowly than P waves.

Scientists use information about where an earthquake occurred and where the P and S waves returned to the surface of the Earth to work out what the different layers of the Earth are, and what they are made from.

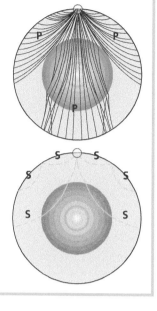

Activity 5: P and S waves Venn diagram

Draw a large Venn diagram on a separate piece of paper. Write the following statements on the Venn diagram to show whether they apply to **P waves**, **S waves** or **both**.

a) Originate at the same point

b) Arrive at seismographs second

c) Travel through solids

d) Cannot travel through liquid outer core

e) Begin at the same time

f) Travel at about 4.5 km per second

g) Can be felt on the Earth's surface

h) Travel at about 8 km per second

i) Cause the first movement you feel in an earthquake

j) Arrive at seismographs first

REMEMBER: HOW TECTONIC PLATES MOVE

After scientists were able to identify the Earth's structure and the temperatures of its different parts, they showed that tectonic plates are moved by **convection currents**.

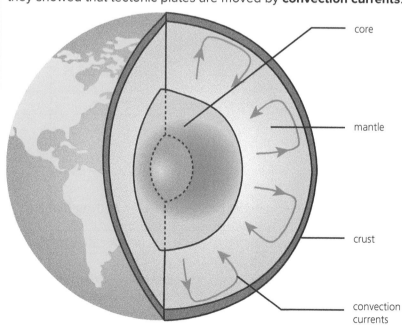

core

mantle

crust

convection currents

Activity 6: Tectonic plates

Learners were asked to write an explanation of how convection currents inside the Earth occur and can move the tectonic plates.

Answer A:

Heat is generated at the Earth's core and passes into the mantle, where convection currents occur. These currents rise to just underneath the Earth's crust and then move along underneath the plates, slowly moving them as they go.

Answer B:

The hottest part of the Earth's structure is the inner core. This heats up the outer core, which in turn heats up the mantle. As the mantle is made up of magma (molten rock), the energy is transferred through it by convection. Convection occurs when thermal energy is transferred through a liquid or a gas by the movement of particles in the substance. The magma closest to the outer core heats up first. The particles in this magma gain more kinetic energy, move faster, and spread apart, becoming less dense than the magma above it. This less dense magma then rises and cooler magma replaces it. The convection currents move beneath the Earth's crust, which moves the tectonic plates.

Answer C:

Different parts of the Earth's structure are at different temperatures. When there are different temperatures then thermal energy will transfer from areas of higher temperature to areas of lower temperature. The outside of the Earth is the hottest as it receives lots of thermal energy from the Sun shining on it every day. This thermal energy therefore heats up the crust on the outside of the Earth. This then heats up the mantle underneath. As the mantle is made of magma (molten rock), then the thermal energy moves through it as convection currents. The convection currents also move the tectonic plates that float on top of the magma.

a) Rank the explanations from best to worst.

....

b) Explain your reasons for your ranking.

....

....

....

....

....

Activity 7: Creating a final concept map

a) Use the concept map word cards that you made in Activity 1. Make sure you now have cards for all the words listed.
b) Lay out the word cards and sort them into groups. Decide on titles for the groups.
c) Place the groups of word cards down on plain paper. When you are happy with them, you can write your titles for the groups above them.

d) Now draw lines to join as many of the cards as you can on the sheet. You can join cards within a group and across different groups. The more lines you make, the more links you are able to show between ideas. This will deepen your understanding and help you make connections and remember key words.

e) Give your concept map to a partner and ask them to:
  – pick two words that you have connected. Explain to them how you have connected these words
  – pick out a sentence or phrase you have written to connect two cards/groups. You must then tell them the two cards/groups that the line connects.

g) You can take photos of your concept map once you have finished it. Use this to help you remember the connections you have made.

TIPS FOR SUCCESS

Go back over the work you have done in this chapter to remind yourself of all the information you have covered. When you are ready, complete this short test.

As you work through it, you can help yourself by:

- reading each question carefully – check you understand the question
- looking for key words to use in your answer
- answering the question in your mind first, before you write it down
- making sure you use correct scientific vocabulary in your answers
- using a piece of spare paper to draft any extended answers first, then when you are happy with it you can write your answer in this book
- checking your answers to make sure that you do not want to make any changes.

Revision test

1 Which of the following are examples of evidence that scientists have used to show there are tectonic plates on the surface of the Earth? Circle the correct answer(s). [4 marks]

Maps of the Earth's landmass   Mapping of rock distribution
Mapping of global temperatures   Mapping of populations
Mapping of earthquakes   Mapping of volcanoes

2 What does monitoring the magnetic polarity of the ocean floor show? Circle the correct answer(s). [2 marks]

A The polarity of the Earth has changed over time.
B The ocean floor is spreading outwards over time.
C The ocean floor is moving together over time.

3 Complete the following table by ticking the correct column to show whether each statement is true or false. [5 marks]

Statement	True	False
The mantle is the outermost layer of the Earth.		
The outer core is the innermost layer of the Earth.		
The outer core is made up of liquid material.		
The crust is the hottest part of the Earth.		
The outer core is the hottest part of the Earth.		

4 Draw lines to match each word to the correct description. [4 marks]

P waves	Measuring instrument used to analyse seismic waves produced in an earthquake
S waves	The two types of waves produced in an earthquake that travel down into the Earth
Seismic waves	Seismic waves that cannot travel through the liquid outer core
Seismometer	Seismic waves that can travel through solids and the liquid outer core

5 Write definitions for each of the following: [3 marks]

a) Magma ....

....

b) Tectonic plate ....

....

c) Seismic wave ....

....

6 Decide if the answer to these questions is **P wave** or **S wave** or **both**. [5 marks]

a) The fastest seismic wave ....

b) The slowest seismic wave ....

c) Felt on the surface of the Earth ....

d) Travels through solids ....

e) Travels through solids and liquid outer core ....

7 Complete the following table by ticking the correct column to show whether each statement is true or false. [5 marks]

Statement	True	False
Convection currents in the crust move tectonic plates.		
Convection currents occur in liquids and gases.		
When liquids are heated, particles in them spread out and the liquid becomes more dense.		
Convection currents in the Earth occur in the inner core.		
Magma that is less dense than the magma above it sinks.		

8 The diagram shows two images of seismic wave patterns. Add labels to indicate which diagram shows P waves and which shows S waves. [2 marks]

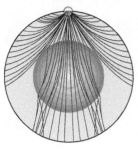 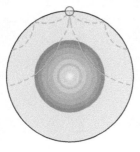

....

9 Draw arrows on the diagram to show the convection currents that move the tectonic plates.

[1 mark]

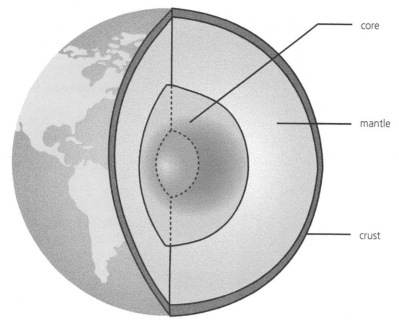

10 Write an extended explanation of how convection currents inside the Earth occur and can move the tectonic plates.

[5 marks]

....

....

....

....

....

....

....

....

....

Chapter 16 Cycles on Earth

REVISION STRATEGY: CYCLES KEY WORD CARDS

For this chapter you are going to make key word cards about cycles on Earth that use words and images. Images are useful and can help you remember things better. You will need to make cards that are about the size of a postcard.

Activity 1: Key word cards

In this activity you are going to make your first key word cards about cycles on Earth. Here is a list of the words that you need to learn and remember related to cycles on Earth.

carbon cycle	decomposition	Industrial Revolution
carbon dioxide	feeding	photosynthesis
climate change	global warming	respiration
combustion	greenhouse gas	

a) Write each word on the front of a card and add any images that will help you remember what the word means.

b) On the back of your card, write a definition. Add as many other ideas as you can that link to the key word.

c) Are there any words you do not know or find it hard to write a definition for? Spend time finding out more about these words. Ask a partner to read your key word cards to check your thinking.

d) Test yourself by only looking at the key word on the front of your card. Can you remember what is written on the back? Ask a partner to test you to see if you can remember everything that you wrote.

REMEMBER: PROCESSES IN THE CARBON CYCLE

There are a number of different processes involved in the cycling of carbon on Earth. These include:

Photosynthesis – This is a chemical reaction that requires energy from sunlight and occurs inside the green chloroplasts inside plant cells. In the chloroplasts, the energy from the sunlight is used to combine water from the soil with carbon dioxide absorbed from the atmosphere. This reaction produces glucose and oxygen. The plant releases the oxygen and the glucose is later converted to starch, which is stored in the leaf.

Respiration – This links to photosynthesis in the carbon cycle. Respiration is a chemical reaction that occurs in the mitochondria of all living cells, releasing energy from glucose. Aerobic respiration uses oxygen and releases energy, as well as carbon dioxide and water.

Feeding – Plants use the carbon they take in during photosynthesis to make carbohydrates, fats and proteins, which they use to keep themselves alive. These substances also form the food for animals, which means that carbon then passes along food chains as animals eat plants and other animals. Food chains link together to form food webs, so there are many paths for carbon to cycle through in the organisms of a habitat.

Decomposition – Carbon in dead plant and animal bodies becomes food for micro-organisms such as bacteria and fungi, and invertebrates such as earthworms and insect larvae. These organisms are called decomposers. As the decomposers feed, they release energy from their food in respiration and produce carbon dioxide, which passes into the air.

Combustion – Humans have used combustion (burning of fuel) for almost half a million years to keep warm and to cook food. When a fuel such as wood is set on fire, the carbon it contains combines with oxygen in the air to produce carbon dioxide and water.

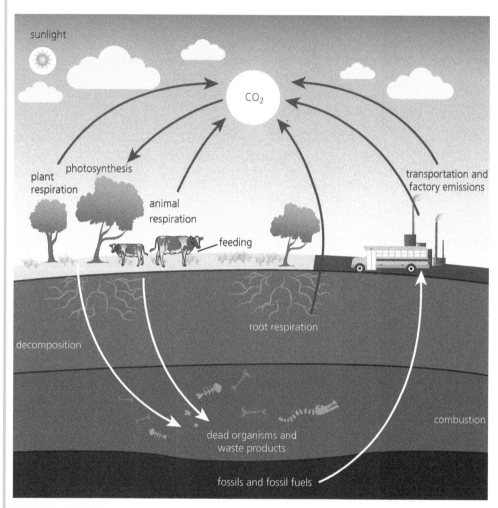

Activity 2: Drawing from memory

Drawing from memory is a useful revision strategy to help you remember key information related to an image. Engaging more of the visual parts of your brain creates a richer experience to help you absorb ideas and remember them better.

a) Look at the image of the carbon cycle in the Remember section above for 10 seconds and then cover it.

b) See how much of the image you can redraw and how many of the 12 labels you can add. Take as long as you want, but do not look at the original image.

c) When you are ready, look at the image for another 10 seconds. Try to remember what you missed and then cover it again.

d) Go back to your drawing and add as much as you can. Take as long as you need, but do not look at the original image!

e) You now have 10 seconds to look at the image for a final time. When your time is up, cover the image.

f) This is your final chance to complete your drawing and include all the details. Take as long as you need.

Activity 3: Remember processes in the carbon cycle

A Stage 9 learner has produced labelled pictures for three different processes involved in the carbon cycle.

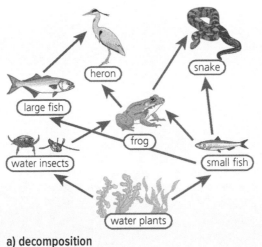

a) decomposition

b) respiration

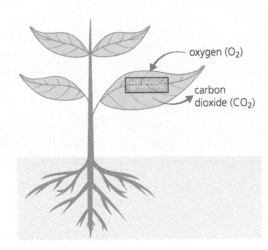

c) feeding

Someone has mixed up the diagrams and their labels. Match each label with the correct diagram.

The label for Picture a) should be ....

The label for Picture b) should be ....

The label for Picture c) should be ....

Activity 4: Climate change

In the table, highlight the correct statement in each row where there is a choice, to explain what climate change is and why it is happening.

Carbon dioxide is classified as a 'greenhouse gas' because ...	
it allows heat energy from the Sun to travel through and reach the surface of the Earth,	it prevents heat energy from the Sun travelling through to reach the surface of the Earth,
but then prevents much of the heat energy radiating from the Earth's surface from passing back out into space, because	and then reflects much of the heat energy back out into space, because
the carbon dioxide absorbs the energy and warms up the atmosphere, like a greenhouse.	
This process is known as global cooling.	This process is known as global warming.
Over two hundred years ago, the Industrial Revolution occurred in several countries and factories were built, which used ...	
coal, oil and gas to power machines using a combustion reaction. This ...	coal, oil and gas to power machines using a respiration reaction. This ...
used up carbon dioxide from the atmosphere.	released more carbon dioxide into the atmosphere.

Scientists have measured global temperatures since 1880 and the patterns show ...		
that the temperature is staying the same.	that the temperature is rising.	that the temperature is going down.

As well as monitoring temperatures, scientists have measured changes over time in	
precipitation, humidity, wind speed and atmospheric pressure.	earthquakes and volcanoes.
These changes are known as climate change and are linked to rises in carbon dioxide levels.	These changes are known as climate change and are linked to falls in carbon dioxide levels.

Activity 5: Climate change evidence

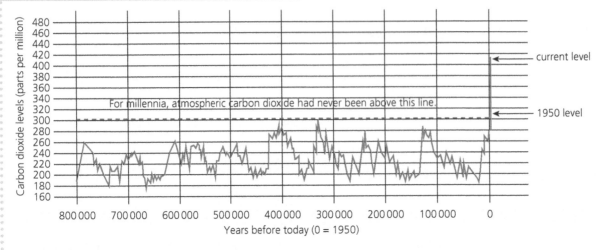

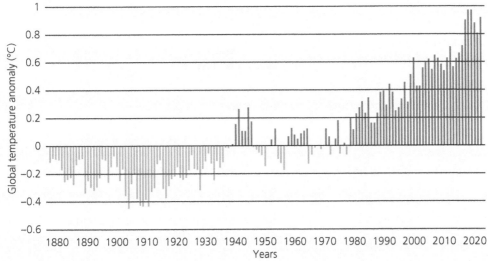

Look carefully at the two graphs, which show data scientists have gathered, and answer the following questions.

a) The temperature graph shows the global temperature anomaly, which means how the temperature changed compared to the long-term average. When did the global temperature anomaly first start to become positive (warmer)? ....

b) According to the data in this graph, for how many years did carbon dioxide levels remain below 300 parts per million? ....

c) In how many years since 1950 have carbon dioxide levels been below 300 parts per million? ....

d) What is the highest carbon dioxide level recorded and when did this happen?

....

e) According to the graph, what is the highest global temperature anomaly recorded and when did this happen? ....

f) When have global temperatures risen at the fastest rate? Explain your answer using data from the graph.

....

....

g) When have carbon dioxide levels risen at the fastest rate? Explain your answer using data from the graph.

....

....

h) Do you think that the rise in global temperature is linked to levels of carbon dioxide in the atmosphere? Explain your answer.

....

....

REMEMBER: IMPACTS OF CLIMATE CHANGE

There are several impacts of climate change observed on Earth:

- As global temperatures rise, ice at the poles, in glaciers and on mountains melts and the water flows into rivers and seas. This causes sea levels to rise and low-lying land and islands can be flooded.
- Many people live along coastlines and in low-lying areas. If these areas are flooded, people will have to move to new places.
- Rising temperatures cause the rate of evaporation from the surface of rivers, lakes and oceans to increase. This means that the air holds more water (humidity), and this disrupts the flow of air around the planet, changing weather patterns almost everywhere around the world.
- New weather patterns can produce extreme weather events, such as very heavy rain leading to flooding, destroying homes and lives, or very dry weather causing droughts and wildfires, damaging crops and livestock.
- Hurricanes (also called cyclones or typhoons) form over warm seas. With rising temperatures, hurricanes will become both more frequent and more severe, because more water will evaporate to form the hurricane. Hurricanes are hugely destructive to life and land through their very strong winds and extreme rain.

REMEMBER: STOPPING CLIMATE CHANGE

There are many ways we can help to stop climate change:

- End the use of fossil fuels and instead use clean, renewable energy sources, such as solar, wind, wave, tidal and geothermal power.
- Use sustainable and less carbon-intensive transport for moving goods and people, and take fewer long trips in order to limit emissions.
- Keep our homes warm by reducing draughts and insulating walls and roofs to reduce the need for heating, and switch away from oil or gas boilers to ground-source heat pumps. If the weather is warm, use air conditioning powered by renewable energy supplies and keep curtains or shutters closed to keep sunlight out.
- Encourage people to reduce or stop meat and dairy consumption to reduce greenhouse gas emissions caused by intensive farming.
- Plant trees or allow land to 're-wild', because photosynthesising plants absorb carbon dioxide. This also includes protecting forests like the Amazon rainforest.
- Protect oceans and the life within them. Large amounts of carbon dioxide are absorbed from the atmosphere into oceans, as it can diffuse into the water or be absorbed by plankton and algae. However, many oceans are overfished, used for oil and gas drilling or threatened by deep-sea mining.
- Reduce how much people consume, as this will mean less processing and transportation of resources on the planet. Reduce our use of plastic, as it is made from oil, and the process of making it is carbon-intense.

Activity 6: Effects of climate change

a) State the **consequences** for each of these **impacts** of global warming:

Rising sea levels ....

....

Changing humidity levels ....

....

Extreme weather events ....

....

b) List all the ways that humans can try to reduce the impact of climate change – these are your **solutions**.

....

....

....

....

....

Activity 7: Fishbone diagram

A fishbone diagram is a way to link many ideas that relate to the same issue. In this activity, you are going to make a fishbone diagram that links ideas about climate change, using your answers from Activity 6. An example of a fishbone diagram is shown below:

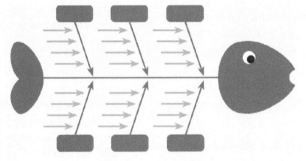

a) Copy the diagram onto a separate piece of paper and write the title 'Climate change' in the head of the fish.
b) On the top bones of the fish, in the boxes at the end of each bone, write the impacts of climate change from Activity 6.
c) Down each 'impact' bone, add the consequences of that impact.
d) On the lower bones of the fish, in the boxes at the end of each bone, write a title for each of your solutions.
e) Along each of your 'solution' bones, add in any ways humans can change what they do to help achieve the solution.
f) Give your diagram to someone else and ask them to test you to see if you can name and link all the ideas about climate change.
g) If you have forgotten anything, draw a picture next to it to help you remember it next time.

Activity 8: Key word card review

Review the list of key words and the cards you made in Activity 1. Add any new learning to the back of your cards. Now test yourself by seeing if you can remember everything that is written on the back of your cards.

Test yourself again in two days' time and see if you can still remember everything. Keep doing this every two days for any cards where you have forgotten details until you can remember everything.

TIPS FOR SUCCESS

Go back over the work you have done in this chapter to remind yourself of all the information you have covered. When you are ready, complete this short test.

As you work through it, you can help yourself by:

- reading each question carefully – check you understand the question
- looking for key words to use in your answer

- answering the question in your mind first, before you write it down
- making sure you use correct scientific vocabulary in your answers
- using a piece of spare paper to draft any extended answers first, then when you are happy with it you can write your answer in this book
- checking your answers to make sure that you do not want to make any changes.

Revision test

1 Which of the following are processes in the carbon cycle? Circle the correct answers. [3 marks]
photosynthesis   fossil fuels   oxygen
carbon dioxide   respiration   feeding

2 Which of the following are mechanisms for absorbing carbon dioxide from the atmosphere? Circle the correct answer(s). [1 mark]
respiration   decomposition   photosynthesis

3 Complete the following table by ticking the correct column to show whether each statement is true or false. [6 marks]

Statement	True	False
Photosynthesis only occurs during daylight hours.		
Respiration only occurs during daylight hours.		
Respiration occurs only in plants.		
Photosynthesis occurs only in plants.		
Burning fossil fuels releases water into the atmosphere.		
The Sun is responsible for global warming.		

4 Draw lines to match the words to the correct sentences. [4 marks]

greenhouse gas	Increase in average temperature of the Earth, which is likely to lead to significant climate changes
carbon dioxide	Any gas in the atmosphere that absorbs some thermal energy from the Sun, with the result that the atmosphere becomes warmer
global warming	Long-term shifts in temperatures and weather patterns
climate change	A colourless, odourless gas consisting of a carbon atom covalently double-bonded to two oxygen atoms

5 Give **two** trends that have been observed in data collected by scientists that provide evidence of global warming caused by human activity. [2 marks]

1 ....

2 ....

6 List **three** different impacts of climate change. [3 marks]

1 ....

2 ....

3 ....

7 Combustion and respiration are both part of the carbon cycle. Decide if each of these descriptions is about **combustion**, **respiration**, or **both**. [7 marks]

a) This process needs oxygen to occur. ....

b) This process occurs in cells. ....

c) This process involves burning. ....

d) This process produces carbon dioxide. ....

e) This process requires a high temperature for it to begin. ....

f) This process produces water. ....

g) This process requires glucose. ....

8 Explain how global warming is affecting sea levels and the impacts this is having. [3 marks]

....

....

....

....

....

....

9 Explain **five** things humans can do to help reduce the problem of climate change. [5 marks]

....

....

....

....

....

....

....

....

....

Chapter 17 Earth in space

CHAPTER INFORMATION

This chapter will help you to revise your learning about the Earth in space. By the end of this chapter you should be able to:

- describe the consequences of asteroid collisions with the Earth, including climate change and mass extinctions
- describe the evidence for the collision theory of the formation of the Moon
- know that nebulae are clouds of dust and gas, and can act as stellar nurseries.

REVISION STRATEGY: HEXAGON MAP

To help you remember the key words of this chapter, you are going to make key word cards. However, for this chapter, your key word cards will be in the shape of a hexagon. As hexagons tessellate, you will be able to fit the cards together to link ideas later in the chapter.

Activity 1: Hexagon key word cards

Here is a list of key words that you need to learn and remember about Earth in space.

asteroid	firestorm	nebula
asteroid belt	impact winter	orbit
Big Splash	main sequence star	protostar
collision theory	mass extinction	satellite
crater	Moon	stellar nurseries
cyclone	near-Earth objects	trojans

a) Write each word on the front of a hexagon-shaped card, adding any images that will help you remember what the word means. On the back of each card, write a definition.

b) Shuffle the cards, place them in a pile and pick up the top card.

c) On a piece of lined paper, write a question where the answer would be the word on the card you have picked. Do the same for the next card on the pile, until you have created a question for every card.

d) Next, lay out your hexagons with the words showing. Ask a partner to read out a question you have written on your lined paper; you then need to find the card that answers the question. You could time how long it takes you to find the answers to all the questions. You can then try this again later in the chapter, and again at the end, to see if you can beat your previous times.

e) Keep your hexagon cards safe, as you will use them again later in this chapter.

REMEMBER: ASTEROIDS

Asteroids are rocky objects mostly found in a ring around the Sun, between the orbits of Mars and Jupiter. This ring is known as the asteroid belt. There are two other groups of asteroids, called trojans. They are found on either side of Jupiter and move in its orbit around the Sun. As the asteroids move in their orbits, they can be affected by contact or non-contact forces, which can cause them to change direction and set up another orbit around the Sun. Some of these orbits may cross the orbits of planets and their moons and, in time, may lead to a collision. Observatories around the world scan the skies for asteroids approaching the Earth (known as near-Earth objects). When one is discovered, the data about it is shared by all observatories worldwide.

Activity 2: What affects the depth and width of a crater?

A group of Stage 9 learners carried out an investigation to answer the question:

How does the height an object is dropped from affect the width and depth of the crater made?

The learners predicted that the greater the height, the greater the final speed at which the object hits and the greater the depth of the crater would be.

The learners dropped a rock into a bowl of sand from different heights. They measured the width and depth of the craters. The data collected by the learners is in the table below.

Height of drop (cm)	Width of crater (mm)	Depth of crater (mm)
10	2	1
20	4	2
30	6	2
40	8	4
50	10	5
60	13	6
70	14	6
80		
90	18	9
100	20	10

a) What is the dependent variable in the investigation? ....

b) What is the independent variable? ....

c) Results that are valid help a scientist to answer the original question they were investigating. What would the learners have to control in their investigation to make the results collected from the experiment valid?

....

....

d) The learners forgot to collect their data for 80 cm. Predict what readings they would have gathered. Explain your reasons.

....

....

e) Using the data, how would you answer the question the learners were investigating? Explain your reasons.

....

....

f) Some of the learners' results do not fit their pattern. What could they have done to increase the reliability of their data?

....

....

REMEMBER: CONSEQUENCES OF ASTEROID COLLISIONS

As well as causing craters, when asteroids collide with Earth there are other consequences, including climate change and mass extinctions.

If the asteroid hits land, it sends a huge amount of dust and ash into the atmosphere. If it lands in the sea, it also sends up water vapour and droplets. These additional materials added to the atmosphere block out the Sun's light and heat, and this leads to rapid climate change. There can be a darkening of the skies and a cooling of the planet's surface. These conditions are sometimes described as an 'impact winter' and they can last for several years.

An impact winter can result in mass extinctions.
- Mass extinctions are events where large numbers of species become extinct.
- They occur when the death rate of many species is greater than the reproduction rate, causing the population to shrink and then die out completely.
- One impact winter occurred 66 million years ago, when scientists believe an asteroid between 11 and 81 kilometres in diameter struck the Earth and made a huge crater. This crater is in Mexico and is called the Chicxulub crater.
- As well as propelling large amounts of dust and ash into the atmosphere, it is believed that the impact would have produced instant firestorms and complete devastation over thousands of square kilometres. The fires would have produced even more ash, all of which led to the impact winter and the mass extinction of many dinosaur species.

Activity 3: Asteroid collisions

a) What are **three** consequences of asteroid collisions on Earth?

1 ....

2 ....

3 ....

b) When an asteroid hits land on Earth, what causes the Earth's climate to change?

....

....

c) When an asteroid hits water on Earth, what causes the Earth's climate to change?

....

....

d) Define the term 'mass extinctions'.

....

REMEMBER: FORMATION OF THE MOON

A satellite is an object that orbits around a larger object in space. Satellites can be artificial and put into orbit, or natural. The Earth is a satellite of the Sun. The Moon is the Earth's only natural satellite. There is a collision theory called the giant-impact hypothesis, or the Big Splash, for how the Moon formed. The theory suggests that about 4.5 billion years ago, the Earth crashed into and destroyed a planet-like object about the size of Mars, which scientists call Theia. This collision resulted in a ring of rocky debris forming around the Earth. In time, forces pulled the rocks in the ring together to form the Moon. Evidence to support this hypothesis comes from a variety of sources.

Activity 4: Evidence and observations

A Stage 9 learner has dropped the notecards they made about the different evidence types and observations that have been used to construct the theory of how the Moon was formed.

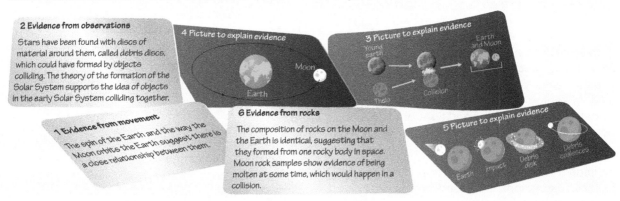

2 Evidence from observations

Stars have been found with discs of material around them, called debris discs, which could have formed by objects colliding. The theory of the formation of the Solar System supports the idea of objects in the early Solar System colliding together.

1 Evidence from movement

The spin of the Earth and the way the Moon orbits the Earth suggest there is a close relationship between them.

6 Evidence from rocks

The composition of rocks on the Moon and the Earth is identical, suggesting that they formed from one rocky body in space. Moon rock samples show evidence of being molten at some time, which would happen in a collision.

Match the notecards back up for the learner. There are three sets with two cards in each set:

1 a title card with a written description to explain the evidence
2 a picture card to help explain the evidence.

Which cards link together to show:

a) evidence from rocks? .... and ....

b) evidence from movement? .... and ....

c) evidence from observations? .... and ....

REMEMBER: STAR FORMATION

There are huge clouds of dust and hydrogen gas spread out across the universe called stellar nebulae. Protostars form as the hydrogen gas comes together in a nebula. The pressure forcing hydrogen atoms together eventually becomes so great that they fuse together, causing hydrogen gas to turn into helium gas and release huge amounts of energy as light and heat. This is when a star begins to shine, and it is then said to be 'born'. The star is now called a main sequence star. Because this 'birthing' process takes place inside nebulae, nebulae are known as stellar nurseries and many stars can be forming inside them at the same time.

Activity 5: How stars are born

You are going to complete a flow chart to describe what happens when a star is born.

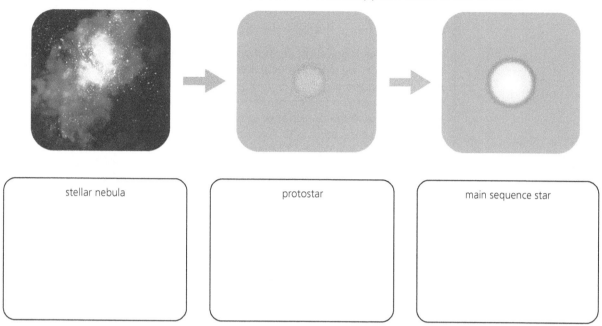

stellar nebula	protostar	main sequence star

In the box under each step of the flow chart, describe what is happening in that step. You must include all the following key terms at some point in the flow chart:

hydrogen gas   gravity   atoms   energy   heat   light
shine   pressure   temperature   helium gas   fuse

Activity 6: Revisit hexagons

a) Use the hexagon cards that you made at the start of this chapter. Create new hexagons for any additional words you would like to add to your hexagon map.

b) Lay out the hexagons with sides touching. You can look back to Chapter 5 for an example of how to lay them out.

c) Say out loud how the hexagon cards that you have placed are connected to each other. For example, you could put the hexagon cards for 'asteroid' and 'crater' next to each other and say that 'When an asteroid collides with a planet or moon, it forms a crater on the surface'. You could then add the hexagon card for 'firestorms' so that all three cards are touching each other. You could then say that 'When a large asteroid hit the surface of the Earth, it produced instant firestorms and left a deep crater'.

d) To challenge yourself, you can:
   – Shuffle the cards and stack them. Take cards from the top of the pile in the order they are stacked and place them in a hexagon map, talking about the links every time you place a card down.
   – Pick out seven cards. Place one card in the centre and the other six cards around it, and explain links between the touching cards.

e) You can take photos of your hexagon map if it helps you to remember the connections you have made.

TIPS FOR SUCCESS

Go back over the work you have done in this chapter to remind yourself of all the information you have covered. When you are ready, complete this short test.

As you work through it, you can help yourself by:

● reading each question carefully – check you understand the question
● looking for key words to use in your answer
● answering the question in your mind first, before you write it down
● making sure you use correct scientific vocabulary in your answers
● using a piece of spare paper to draft any extended answers first, then when you are happy with it you can write your answer in this book
● checking your answers to make sure that you do not want to make any changes.

Revision test

1 Which of the following are locations of asteroids? Circle the correct answer(s).   [2 marks]
   Between the Sun and Earth   Between Earth and Mars   Between Mars and Jupiter
   Between Jupiter and Saturn   In the same orbit as Jupiter

2 Which of the following statements are correct? Circle the correct answer(s).   [2 marks]
   A Asteroid collisions can cause craters.
   B Asteroid collisions can cause climate change.
   C Asteroid collisions can cause global warming.

3 Complete the following table by ticking the correct column to show whether each statement is true or false.   [6 marks]

Statement	True	False
Asteroids that hit land on Earth send dust and water vapour into the atmosphere.		
Near-Earth objects are satellites approaching the Earth.		
Dust, ash clouds, water vapour and droplets block out light from the Sun.		
Darker skies after asteroid collisions result in a warmer planet.		
Mass extinctions occur when death rates are higher than birth rates.		
Big enough collisions from asteroids can produce fire hurricanes.		

4 Draw lines to match each word below to the correct description.   [4 marks]

Moon	A collision theory hypothesising how the Moon formed
satellite	The Earth's only natural satellite
debris disc	A moon, planet or machine that orbits a planet or star
Big Splash	A ring of rocky material orbiting a star or planet

5 Write definitions for each of the following phrases: [2 marks]

   a) Impact winter ....

   ....

   b) Mass extinctions ....

   ....

6 State three different pieces of evidence that support the collision theory for how the Moon was formed. [3 marks]

1 ....

2 ....

3 ....

7 Decide if the answer to these questions is **asteroid** or **moon** or **both**. [6 marks]

a) Made from rock material ....

b) Cause collisions ....

c) Formed from a collision ....

d) Orbits the Earth ....

e) Orbits the Sun ....

f) Move around the Sun. ....

8 Put the following pictures into the correct order to show how a star is born. [3 marks]

A: main sequence star

B: stellar nebula

C: protostar

The correct order is: ......, ..., ...

9 Explain how stars are born, using all the words below in your paragraph. [5 marks]

| hydrogen gas | gravity | atoms | energy | heat | light |
| shine | pressure | temperature | helium gas | fuse | |

....

....

....

....

....

....

....

....

....

....

....